RURAL LIGHTING

RURAL LIGHTING
A guide for development workers

Jean-Paul Louineau, Modibo Dicko,
Peter Fraenkel, Roy Barlow
and Varis Bokalders

Practical
ACTION
PUBLISHING

Practical Action Publishing Ltd
25 Albert Street, Rugby,
Warwickshire, CV21 2SD, UK
www.practicalactionpublishing.com

A catalogue record for this book is available from the British Library & Library of Congress

ISBN 978-1-85339-200-9 Paperback
ISBN 978-1-78044-582-3 Digital book

Citation: Louineau, J. (1994) *Rural Lighting: A guide for development workers*, Rugby, UK: Practical
Action Publishing https://doi.org/10.3362/9781780445823

Since 1974, Practical Action Publishing has published and disseminated books and information in
support of international development work throughout the world. All print editions are produced
and distributed via ethical and sustainable print on demand global facilities.

Practical Action Publishing is a trading name of Practical Action Publishing Ltd (Company Reg.
No. 01159018 | VAT 880 9924 76). All profits are covenanted back to its parent group, Practical
Action (Charity Reg. No. 247257).

The manufacturer's authorised representative in the EU for product safety is Lightning Source
France, 1 Av. Johannes Gutenberg, 78310 Maurepas, France. compliance@lightningsource.fr

Contents

CONTENTS

Foreword

Access to lighting in homes, public institutions and other places means that the time available for productive use can be extended beyond the period of daylight. It makes it easier to study, to work, to take care of your children, etc. Lighting is an essential part of the development process, to raise the quality of daily life in rural areas. In many places there is no access to an electric grid, and rural electrification is a very expensive concept. This book describes other possibilities to get lighting in such areas. The initiative to produce this book came from the Stockholm Environment Institute, and its publication is part of our co-operative programme with SIDA on 'Energy, Environment and Development'.

In our work with energy and developing countries we soon realized that small energy problems can be as important as big ones. A little bit of useful energy in a remote area may not show up in the statistics but can make life much easier for the people concerned. The energy might be needed for pumping, milling, lighting or other everyday needs. Our colleagues from the Swedish Mission Council, working in the field, agreed with us that this was an important issue. So that is why we decided to write this book.

The idea of the book is to give you the know-how necessary to be able to understand the characteristics and technologies available for lighting. We outline a methodology on how to make an informed choice of lighting system, and finally how to go about it. In order to write this book we co-operated with IT Power in the U.K., an organization which has a lot of experience in lighting systems in remote areas.

We hope that this book will help people to spread light in dark places.

Lars Kristoferson
Stockholm Environment Institute
November 1993

Preface

This guide book is a result of a co-operative project involving The Stockholm Environment Institute (SEI), IT Power Ltd, and the Swedish Mission Council (SMC). It has been sponsored by the Swedish International Development Authority through SEI. The Stockholm Environment Institute runs an information programme on renewable energy for development which has resulted in a series of publications and seminars. This book is one of the results of this work.

IT Power has substantial experience in the field of lighting systems in remote areas in developing countries, for example with the installation and maintenance of 750 photovoltaic lighting systems in health centres in Zaire. IT Power works more broadly in all areas of renewable energy application, including lighting.

The origin of the book derives from the needs of SEI and SMC field staff and other development workers worldwide who have found much of the information currently available on lighting systems to be fragmented, scarce, too theoretical, and therefore difficult to use.

This book, therefore, aims to provide information both on lighting needs in general and on the technologies, performance and economics of various available lighting systems.

The objective is to assist development workers and project managers working in rural or semi-urban development, health, farming and small business sectors to decide:

- What level of light is necessary to meet the needs;

- Which type of lighting sources would be most suitable;

- How to implement cost-effective lighting systems, including:
  - ✓ enhancing light sources
  - ✓ assessing the energy and balance of system requirements
  - ✓ sizing and specifying
  - ✓ working out the economics
  - ✓ procurement and installation
  - ✓ maintenance and operation.

This is the fourth in the series of guides for development workers on energy technologies for sustainable development. Other guides in this series are:

Solar Photovoltaic Products
First ed. 989 (by A. Derrick *et al*)
Updated 1991
ISBN 1 85339 091 7

Micro-hydro Power
1992 (by P.L. Fraenkel *et al*)
ISBN 1 85339 029 1

Windpumps
1993 (by P. L. Fraenkel *et al*)
ISBN 1 85339 126 3

Contact persons are:

Peter Fraenkel	Varis Bokalders	Karl-Erik Lundgren
IT Power Ltd	Stockholm Environment Institute	Swedish Mission Council
The Warren	Box 2142	Office for international
Bramshill Road	103 14 Stockholm	Development Co-operation
Eversley	Sweden	Tegnergatan 34 n.b.
Hampshire, RG27 0PR		113 86 Stockholm
UK		Sweden
Fax +44 734 730820	Fax +46 8 723 0348	Fax +46 8 31 58 28

Acknowledgements

The authors wish to thank the Swedish International Development Authority, SIDA, for supporting the production of this guide and providing the finance for research work.

The review comments from Lee Schipper and Paul Konor are gratefully acknowledged, as are the comments and suggestions received from colleagues of the authors.

The guide could not have been produced without the kind assistance from the international lighting industry, both electrical and flame-based , and the photovoltaic manufacturers or suppliers in providing data and some of the photographs.

The authors wish also to thank Kathleen Glancey for the desktop publishing of the text of this book.

NOTICE

Introduction

1

Wisdom excelleth folly, as far as light excelleth darkness...

Ecclesiastes

1.1 Lighting

Lighting is considered a universal need. Similarly it is argued that it is a determinant of quality of life. While this may not be true in all cases, it can certainly be said that lighting has the capability to transform lifestyles.

In developing countries, people have access to lighting but often of poor quality and insufficient quantity. Furthermore, lighting can represent a high cost relative to household income. A large share of the non-food budget can be absorbed in the purchase of fuels for lighting.

As in industrialized countries, rural and semi-urban people in developing countries consider lighting to be of great value. Lighting can add many productive hours to the work day. In the home, sufficient lighting can facilitate activities such as studying, handicrafts and other productive activities such as seed sorting or sewing.

Commercially, lighting can increase business hours or improve the attractiveness of available services. In schools and educational institutions, lighting can increase study time and allow for evening/ adult classes or training. In rural dispensaries and hospitals, lighting can improve the quality of nighttime care and allow for emergency procedures to be carried out at night.

1.2 About this Book

In general, evening activities which may be common in urban, electrified areas are curtailed in more rural areas due to lack of light. It is difficult however to quantify or qualify the economic and social value of lighting. This is because it usually only indirectly facilitates

Figure 1.1 Photovoltaic lighting installation in a classroom of the Kilwa Medical Technical institute in Shaba, Zaire. Nurses in training use this room for study in the evenings

economic activities and development. For this reason it is an area that is often neglected or regarded as a luxury rather than a necessity by planners. Hence, the literature on development technology for clean water, sanitation or agriculture is copious by comparison with the literature on lighting.

This book is hoping to fill this gap. Its objective is to review the options for providing lighting in rural areas in some detail. It simplifies the procedures for specifying the most cost-effective systems. It also guides the reader in understanding and interpreting manufacturers' specifications for their products.

Since the main reason for the lack of good lighting in rural areas is the absence or inaccessibility of grid electricity, a major element of this book is to review the options for producing and storing the electricity needed for lighting. Non-electric lighting possibilities are also discussed, as some of them are in widespread use. New developments in the field of solar photovoltaics are also presented as these provide good quality lighting systems independently of the mains and are becoming increasingly available.

Another important topic covered in detail is the use of primary or secondary cells for torches and handlamps. Primary cells (otherwise known as 'dry cells') may seem relatively inexpensive per unit but are simply thrown away when exhausted while secondary cells (usually known as rechargeable cells) can be re-used many times, so they work out cheaper in the long run.

Although this book is primarily aimed at showing methods for minimizing the energy needs for providing light where there is no mains electricity supply, it is hoped that aspects of this book on the selection and use of efficient lighting will be of considerable relevance when planning lighting even in conventional urban applications. It is also important to find ways of improving the cost-effectiveness of lighting and to minimize the electricity requirements in this period of increasing concern about the environmental impact of excessive energy consumption patterns.

The chapters can be read independently so as to allow the reader to have quick access to relevant and practical information. The reader may use only some parts of the book for reference or perhaps only the Buyers Guide on lighting products. Others, who may be planning complex and elaborate lighting arrangements (perhaps for businesses, health centres or hospitals rather than for single domestic residences), will be able to review their requirements in a more analytical way, as explained in the appropriate chapters.

The authors hope this book will 'throw some light' on an important but seriously neglected topic. It is planned that the book will be revised from time to time in the future to keep the technical contents and the Buyers Guide as up-to-date as possible.

Light and Lighting 2

2.1 Introduction

What is light? How can it be quantified and measured? What are the different sources of light? What are the required light levels for different interiors and activities? What are the relations between energy and light? These are some of the important questions with which this chapter will deal. Since this is a rather technical chapter intended for reference purposes, the reader with only a general interest might consider going straight to Chapter 3.

2.2 Visible Radiation and Light

When a piece of iron is heated in a fire, it first becomes red and emits both some reddish light and some heat. If the fire is very strong and heats the metal up to a high enough temperature, it becomes yellow and eventually white and produces a bright white light and a great deal of heat (e.g. a blacksmith sweats not only because of his physical exertion but also because he receives the heat

and light from the metal). Light, like heat, is a form of energy.

In general, when a metallic material becomes hot, it emits radiation. This happens in an incandescent bulb when it is switched on. The thin filament is heated to such a high temperature by the electrical energy fed into it that it begins to glow and emit radiation; some of this radiation is invisible (i.e. heat radiation) and some visible (i.e. light).

Light is electromagnetic radiation, a form of energy that can radiate from a source through air or space. Electromagnetic radiation is a physical phenomenon which behaves according to fixed laws. Each type of radiation is characterized by its wavelength. The electromagnetic spectrum contains different types of radiation ranging from X rays with short wavelengths, ultra-violet radiation, **visible radiation**, infrared 'heat radiation' (wavelengths ranging from 1 to 10 micrometres), microwaves and finally radio waves.

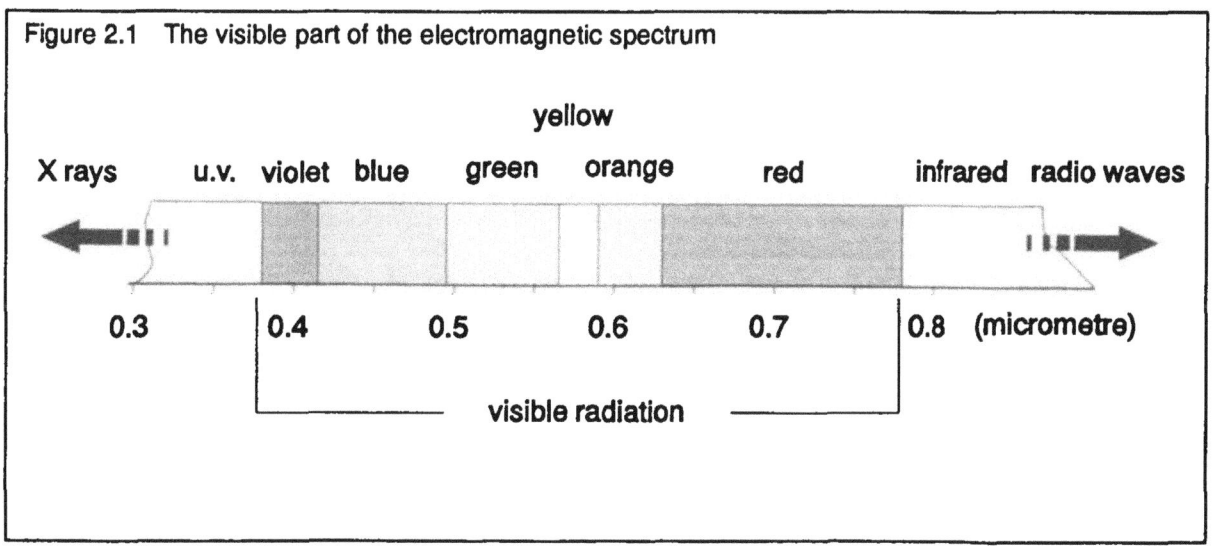

Figure 2.1 The visible part of the electromagnetic spectrum

The human eye and radiation

The human eye is sensitive only to radiation with wavelengths between 0.38 and 0.76 micrometre (one micrometre is 10^{-6}m or one millionth of a metre). Figure 2.1 illustrates the visible spectrum, which appears to the eye as the familiar colour bands of a rainbow, ranging from deep red and merging through shades of red, orange and yellow, through greens and blues, to violet as the wavelength shortens.

Light of one fixed wavelength is known as 'monochromatic' light and has a pure colour (e.g. the yellow light of a low-pressure sodium lamp). Human sight is, however, incapable of distinguishing between all these wavelengths as separate colours, but combines the above named groups into a certain **colour impression**. When all the colours are seen simultaneously, as in sunlight and artificial light, they appear to the eye as white.

Visible radiation causes the impression of both light and colour on the human eye. Table 2.1 illustrates the relationship between wavelength and colour. The effect of radiation on the human eye varies with wavelength. The eye is most sensitive to radiation with wavelengths situated in the middle of the visible spectrum (i.e. yellow-green at a wavelength of 0.555 micrometre). Towards the boundaries of the visible spectrum (red and violet radiation) the effect decreases gradually, until they cease to appear visible. This variation is known as the **spectral sensitivity** of the human eye and has been determined by means of measurements on a large number of test subjects.

Table 2.1	Relationship between wavelength and colour
Physical wavelength (micrometre)	Physiological colour impression on the eye
0.380 to 0.420	violet
0.420 to 0.495	blue
0.495 to 0.566	green
0.566 to 0.589	yellow
0.589 to 0.627	orange
0.627 to 0.760	red

In conclusion, **light** is visible radiation as evaluated by the human eye: the radiation is the cause, the impression of light is the result.

2.3 Sources of Light

Sources of light can be classified into two fundamental categories: natural light and artificial light. Natural light is not easily controllable, while artificial light can be fully controlled and adjusted to give different effects.

Natural light

The main characteristic of natural light is its variability: its level varies constantly with changing meteorological conditions, time of the day, seasons, and specific location on the Earth.

Natural light can be divided into two components: sunlight and skylight. The incoming light from the sun is partially absorbed and scattered when it reaches the Earth's atmosphere. Sunlight is the light received directly from the sun. Skylight is the scattered light received from the other luminous parts of the sky. **Daylight** is the combination of skylight with any available sunlight. Sunlight tends to have a warmer (i.e. more yellow) colour than skylight, which is relatively bluish.

Natural light (i.e. daylight) can and should be used whenever possible as a light source in buildings (see Chapter 5).

Artificial light

Artificial light sources fall into two main categories, flame-based and electrical light sources.

- **Flame-based light** sources have been used since the discovery of fire; they evolved from firelight, oil lamps and candles to the modern kerosene and gas lamps, which are still in widespread use (see Chapter 3).

- **Electrical light sources** appeared very recently in history, towards the end of the last century, but are now the norm for modern lighting and are also much more energy efficient than flame-based light sources. They can be divided into two broad classes on the basis of their operating principles: incandescent lamps and discharge lamps (see Chapter 4).

2.4 Light and Energy

A source of light needs energy, either electrical (e.g. bulbs) or heat (e.g. flames). **Energy** is required for lighting just as for doing any kind of work (e.g. lifting water or cooking food). The rate at which it is used is measured as **power**. A certain amount of work can be done slowly, using little power or quickly using more power, involving the same amount of energy in either case.

Energy = power x time

The SI unit for energy is the Joule (J), while the power is expressed in Watts (W) and the time in seconds. Energy can also be expressed in Wh when the power is given in Watts and the time in hours. As a consequence, 1Wh is equal to 3600J.

Energy is usually expensive and finite. For sustainability reasons and for energy conservation, it is essential to understand the relationship between energy input and light output to be able to choose an efficient and appropriate lighting source.

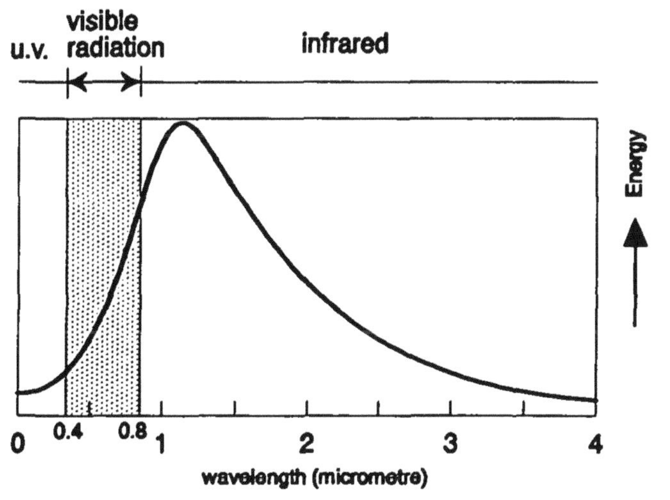

Input: energy consumed

It is fairly simple to evaluate the quantity of energy consumed by a light source (i.e. energy input). For the flame-based systems it is necessary to know the net heating value (see Chapter 6) and the rate of consumption of the fuels used or for electrical lighting systems, the power rating of the lamps.

Example of energy consumption for an electrical lighting source

A 100W electric bulb running for two hours will consume 100W x 2h = 200Wh of energy. A 200W electrical bulb running for only one hour will consume the same amount of energy (i.e. 200W x 1h = 200Wh).

Example of energy consumption for a flame-based lighting source

The power of a flame-based source is equal to the fuel consumption per second multiplied by the net heating value of the fuel (i.e. fuel energy content, see also Chapter 6). For example, a kerosene lantern burning 0.04 litre/hour (l/h) of kerosene (kerosene net heating value is 36,000,000J/l or 36MJ/l). Hence, it has a power of 0.04 x 36,000,000 / 3600 = 400W. Operating for two hours, it will consume 400W x 2h = 800Wh.

Output energy: light

It is also possible to measure the quantity of energy that is contained in the output of a lighting system, i.e. in the light, but this needs some explanation using an example.

It is possible to measure how the emitted energy of a 100W incandescent lamp is distributed over the radiation spectrum (see Figure 2.2). It can be seen that an incandescent

Figure 2.2 Radiation spectrum of a 100W incandescent lamp

bulb emits a great deal of the transmitted energy in the infrared range and only a small part as visible light. Overall, the energy, which is the area between the curve and the horizontal axis, emitted in the visible range accounts for only 10% of the input electrical energy. The rest is radiated in the infrared as heat. However, this 10W is distributed over the visible spectrum, and the eye will react more strongly to some wavelengths than to others. Thus 1W of red light will appear dimmer than 1W of yellow or green light.

Light energy as evaluated by the human eye is not expressed in watts but in **lumen** or less often in light-watts. These measurements take the spectral sensitivity of the eye into account. For example, the total light energy emitted by a 100W bulb is 1700 lumen (or 2.5 light-watts) which is only a fraction of the

---

How much light does a light source produce?

In order to find out how much light an incandescent bulb gives, the emitted energy can be expressed in **lumen** or in light-watt (i.e. 1 light-watt is 680 lumen). One light-watt is defined as one watt of visible radiation emitted at a wavelength of 0.555µm; this is because the eye sensitivity is 100% at this wavelength (see Figure 2.4). If a 100W incandescent bulb emitted all of its light at 0.555µm, the 10W of energy in the visible part of the spectrum would be equal to 10 light-watts. The spectral energy distribution in Figure 2.3 shows that this is not the case. The 10 watts are distributed over the whole visible range.

The number of light-watts can be found by multiplying the energy emitted in every wavelength in the visible range by the relative luminous efficiency curve which takes into account the sensitivity of the eye. For example, if an output of 1.5W is emitted at 0.6 micrometre having a luminous efficiency of 63% (see Figure 2.4), the light-energy is equal to 1.5 x 0.63 = 0.94 light-watt (see Figure 2.5). When these products have been calculated for every wavelength, the total number of light-watts is determined by adding all the results (summing the area under the curve). For example, the total light-energy emitted by a 100W bulb is about 2.5 light-watts or 1700 lumen.

Figure 2.3   Visible radiation spectrum from a 100W incandescent lamp

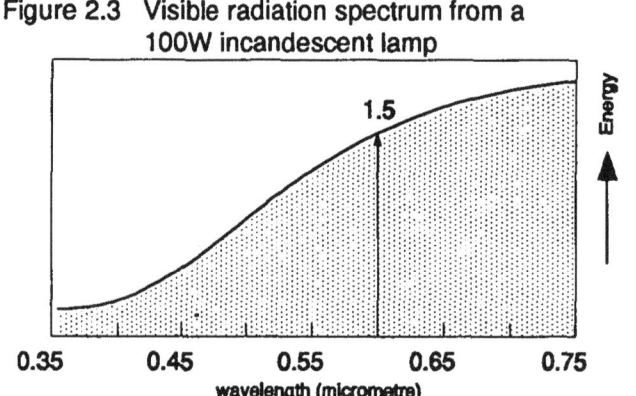

Figure 2.4   Relative luminous efficiency curve

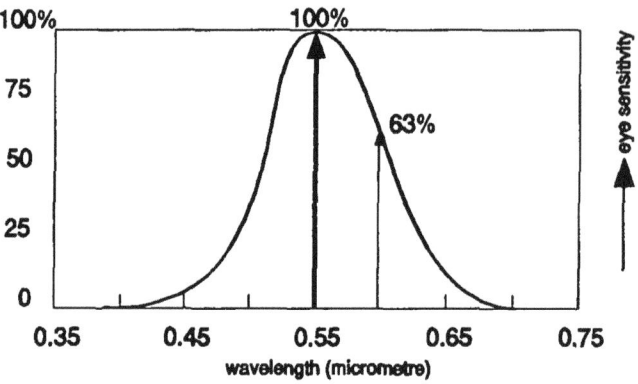

Figure 2.5   Light-energy wavelength
(the grey area represents the quantity of light as seen by the human eye)

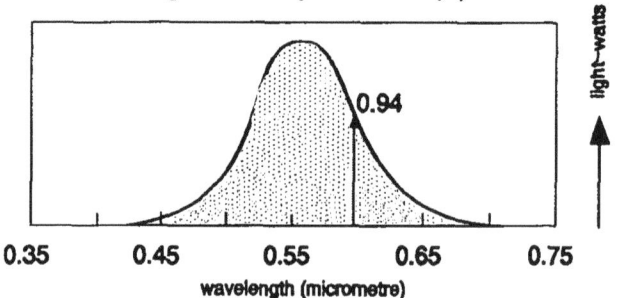

10W of energy in the visible range. This quantity is called by definition the **luminous flux**. The way this 1700 lumen is obtained is explained as clearly as possible for the reader in the box 'How much light does a light source produce?'.

2.5 Photometric Quantities

This section is included as a means for the reader to interpret the different technical terms and units that commonly appear in specifications relating to lamps and to understand how photometric quantities are related or even useful. The quantities 'luminous flux', 'luminous efficacy' and 'illuminance' are very important and should be well understood.

Definitions

Illuminance (E) [unit lux (lx)]

Illuminance is a measure of how well an illuminated object is lit. The illuminance is the luminous flux falling on unit area of a surface, and is expressed in lux (i.e. 1 lux = 1 lumen/m²).

Examples of illuminances (approximate values)	
Summer, midday, in an open field, clear sky	90 000 lx
Summer, midday, under a balcony	3000 lx
Full moon and bright sky at night	0.50 lx
Office with good lighting	300 lx
Minimum for reading	25 lx
Object lit by:	
a candle at 0.5m	6 lx
a kerosene lamp at 0.5m	25 lx
a solar portable lamp (5W fluorescent tube) at 0.5m	60 lx
a 100W incandescent bulb at 0.5m	500 lx

Luminous intensity (I) [unit candela (cd)]

The **luminous intensity** of a light source is a measure of how much light is emitted in a particular direction. By definition, it is the luminous flux per unit of solid angle (or steradian) in a given direction. It is expressed in candela (cd).

Examples of luminous intensity straight ahead (approximate values)	
40W/240V Spot incandescent lamp (80° beam angle)	120 cd
40W/240V Spot incandescent lamp (35° beam angle)	400 cd
35W/12V halogen incandescent spot (10° beam angle)	8000 cd

Luminous flux (Iu) [unit lumen (lm)]

The **luminous flux** is the total amount of light emitted by a source, or received by a surface. The quantity is derived from the radiation spectrum of a light source by evaluating the radiation in accordance with the spectral sensitivity of the standard eye as explained above. It is expressed in lumen (lm).

Examples of luminous flux	
Candle	5 lm
Kerosene wick lamp	100 lm
100W incandescent lamp	1700 lm
60W fluorescent tube	3600 lm
70W low-pressure sodium lamp	11 000 lm

Luminance [unit cd/m²]

The **luminance** expresses how bright a source appears. It is the luminous intensity per cm² or m² of surface area of a light source or of an illuminated area. The SI unit for luminance is the candela per square metre (cd/m²). Luminance used to be called brightness. For example, the surface of a fluorescent tube is less bright to look at than the surface of an incandescent bulb (i.e. 0.7cd/cm² against 700cd/cm² as can be seen below).

Examples of luminance	
Sun	165 000 cd/cm²
Filament of an incandescent lamp	700 cd/cm²
Low-pressure sodium lamp	20 cd/cm²
Fluorescent tube	0.7 cd/cm²

Luminous efficacy (I) [unit lm/W]

The energy-to-light conversion, called by definition the **luminous efficacy**, is given in lumen/watt (i.e. energy-light output divided by the input power of the source). In the case illustrated above, it is equal to 17 lm/W (i.e. 1700 lumen / 100 Watts).

Relations between photometric quantities

If a small light source has a total luminous intensity of one candela, then the luminous flux transmitted to the surface of a sphere of one metre radius surrounding the point source of light will be one lumen for each square metre of the sphere surface. This relation is not of much direct use, but it summarizes the relationship between candelas and lumens.

Average illuminance on an area

This is probably the most important and practical relation to know as it will be extensively used in Chapter 9 on sizing lighting systems. The average illuminance on an area is equal to the total luminous flux on the area, divided by the area.

$$E = lu / S$$

where:

E = average illuminance (lx);

lu = average luminous flux (lm);

S = surface to be lit (m²).

For example, when an incandescent bulb of 500 lm lights an area of 10m², the average illuminance is 50 lx. This example assumes

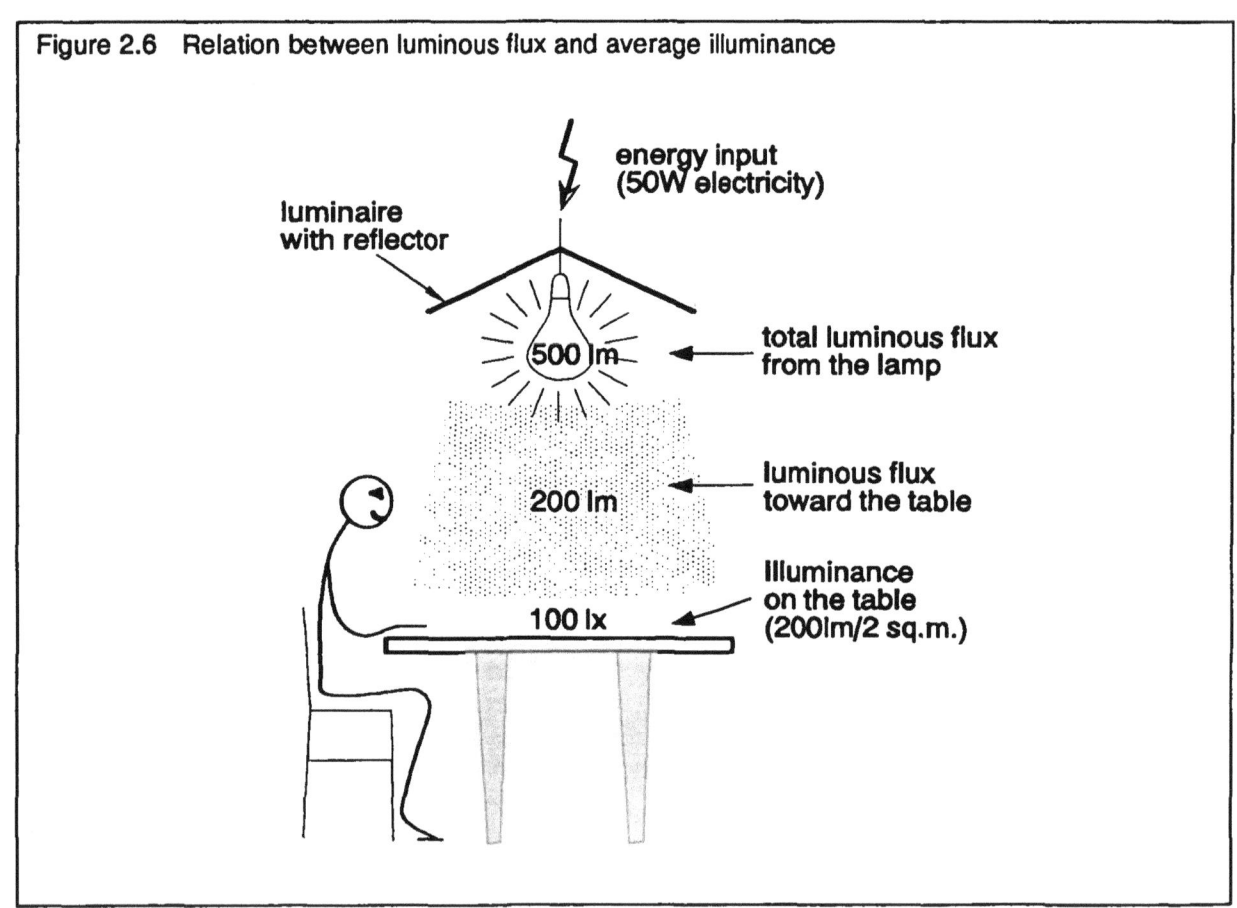

Figure 2.6 Relation between luminous flux and average illuminance

energy input
(50W electricity)

luminaire
with reflector

500 lm

total luminous flux
from the lamp

200 lm

luminous flux
toward the table

100 lx

Illuminance
on the table
(200lm/2 sq.m.)

that all the light has reached the area. If only 200 lm reaches a working area of 2m², as illustrated in Figure 2.6, the average illuminance is 100 lx.

Illuminance at a point

The illuminance at a point can be calculated easily when the size of the source is small compared with its distance from the surface. A point source could be a bulb or even a small fluorescent tube. The following formula is used to calculate illuminance at a point:

$$E_{point} = (Li / d^2) \times cosine\ b$$

where:

E_{point} = illuminance at a point (lx);

Li = luminous intensity (cd);

d = distance between the source of light and the point (m);

b = angle between the beam and a line perpendicular to plane of the lit surface(°).

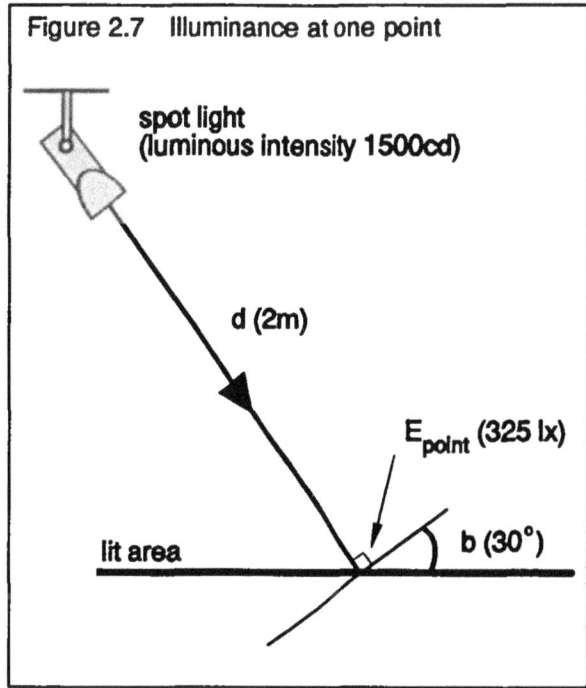

Figure 2.7   Illuminance at one point

spot light
(luminous intensity 1500cd)

d (2m)

E_point (325 lx)

lit area   b (30°)

For example, Figure 2.7 demonstrates a spot light of 1500cd at the centre of the beam. It gives, on a surface perpendicular to the beam and at 1m, an illuminance of 1500/1² x 1 = 1500 lx and at 2m, 1500 / 2² x 1 = 375 lx. If the surface is turned through an angle of 30° from being square on to the light beam, the illuminance at 2m will be 375 x 0.866 = 325 lx (0.866 = cos 30°).

Illuminance at a point using a polar curve

The luminous flux from a light source is never emitted in a completely uniform manner. A characteristic for the light distribution of a lamp or a luminaire can be found from its polar intensity diagram, usually illustrated in the manufacturer's catalogue. Without going into details, this curve gives in each direction, and for a luminous flux of usually 1000 lm, the luminous intensity of the light in candela. This curve is useful in determining the illuminance at a point using the formula '$E_{point} = Li / d^2$'. For example, in Figure 2.8, the

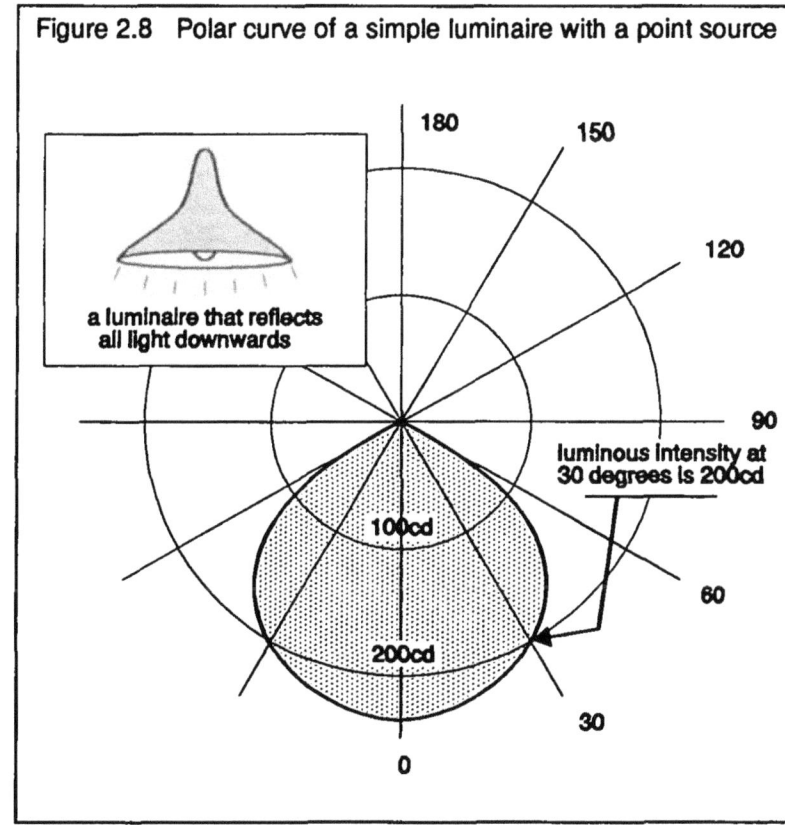

Figure 2.8   Polar curve of a simple luminaire with a point source

a luminaire that reflects all light downwards

180   150

120

90

luminous intensity at 30 degrees is 200cd

100cd

60

200cd

30

0

polar curve of a simple luminaire shows that at 30°, the luminous intensity is 200cd. Hence at 1.5 metre and 30° and with a lamp of 1000 lm, the illuminance is $200/1.5^2 = 88$ lx. If the lamp has a luminous flux of 2000 lm, the illuminance would be double.

When a polar curve is not available and in the special case of uniform output from a light source in all directions, it is possible to calculate the luminous intensity from the lumen output using the formula:

$$Li = lu / (4 \times 3.14)$$

Where:

Li   = Luminous intensity (cd);

lu   = Luminous flux (lm);

3.14 = pi.

For example, an incandescent halogen bulb, without a reflector, of 1200 lm and which is very compact (i.e. approximately a point source emitting light in all directions), has a luminous intensity of 1200 lm / (4 x 3.14) = 95cd in every direction. By knowing the luminous intensity, illuminance at one point can be calculated as has been shown earlier.

2.6 Colour and Colorimetry

Three attributes are used to describe colour: the hue (e.g. red, orange, green, blue, etc.); the value (of a colour) which describes the brightness for luminous objects and the chroma (of a colour) i.e. its strength or depth. It is beyond the scope of this book to review in detail how colours are best defined, but it is important to explain how a light source affects the colours of objects that it illuminates.

A distinction is made between the **colour temperature** (i.e. appearance) of a light source when you look at the light itself and the **colour rendering** that it gives to surfaces when it shines onto them.

Colour temperature

The precise colour appearance of a light source can be defined in practice by the **correlated colour temperature (CCT)**. The CCT of common white light sources ranges from approximately 2700 to 6500 degrees Kelvin (°K). Temperature in °K is equal to the temperature in degrees Centigrade (°C) + 273.

A 2700°K source is seen as 'reddish' and having a warm appearance, hence it can be used to create a warm atmosphere. A 6500°K source is seen as 'bluish' and having a cold appearance. It can be somewhat confusing that a lower CCT gives a warmer appearance and a higher CCT a cooler one. Table 2.2 shows the three basic CCT classes adopted by the Commission Internationale de l'Eclairage (CIE) for lamp classification.

The CCT is not related in any way to the running temperature of a lamp, nor it is a guide to its colour rendering properties.

Colour rendering index

The colour of an object will appear different under different types of light. The **colour rendering index** expresses how a light source compares with natural light or daylight in its

Table 2.2 Correlated colour temperature		
Correlated colour temperature (CCT)	CCT class	Visual appearance of the light source
CCT < 3300°K	Warm	Reddish
3300 < CCT < 5300°K	Intermediate	Intermediate
5300°K < CCT	Cold	Bluish

Table 2.3 Colour rendering groups

CIE Colour rendering %	Typical applications	Example of lamps
90 (Good)	Accurate colour matching	Incandescent lamps
80-90 (Good)	Accurate colour judgement or good colour rendering for reasons of appearance	Fluorescent tubes (with triphosphor fluorescent coating)
60-80 (Medium)	Moderate colour rendering	Standard Fluorescent tubes
40-60 (Medium)	Little significant colour rendering, but marked distortion of colours unacceptable	High-pressure mercury lamps
20-40 (Poor)	Colour rendering not important and colour distortion acceptable	Low-pressure sodium lamps

ability to make objects appear in their natural colours. Put more precisely, it is a measure of the degree to which the colours of surfaces illuminated by a given light source conform to those of the same surfaces under a reference light. Some form of daylight is taken as the reference source. Perfect agreement between the source being judged and the reference source is given a value of 100. Table 2.3 shows the different colour rendering groups adopted by the CIE.

Colour rendering is a very important factor where artificial lighting is used in situations demanding accurate perception of colours, such as for surgical work, artwork or certain handicraft activities.

2.7 Flicker, Stroboscopic Effects and Glare

AC electrical power supplies induce in all types of lamp an inherent fluctuation in light output. Most often this is not visible, but when it becomes visible, it is called flicker. Flicker may cause people some discomfort and distraction.

Stroboscopic effects occur where lighting flicker is at a similar frequency to some movement of machinery or other equipment. This causes the kind of phenomenon where, for example, the spoked wheels of a stage coach in a movie appear to be rotating backwards because the movie camera shutter

Table 2.4 Summary of lighting quantities

Quantity	Informal description	Unit
Illuminance	how well an object is lit	lx
Luminous flux	how much light is emitted by a source	lm
Luminous efficacy	how energy efficient is a light source	lm/W
Luminous intensity	how much light is emitted in one direction	cd
Luminance	how bright is a source	cd/m^2
Colour rendering	how a light source compares with natural light	%
Colour temperature	how a light source appears (e.g. warm or cold)	°K

works at a speed near to, but not exactly the same as the speed of rotation of the spokes of the wheels. This can in certain circumstances cause confusion or difficulty for a viewer.

Glare occurs when some parts of the field of view have illuminances much greater than the average illuminance in an interior. Two types of glare are distinguished: discomfort glare, resulting in a physical discomfort and disability glare, which results in a loss in visual performance.

Glare can also be categorized into direct and indirect glare. Direct glare occurs when very bright sources of light (for example, luminaires or windows) shine directly into the eyes. Indirect glare is due to light reflected into the eyes from shiny surfaces in the range of vision.

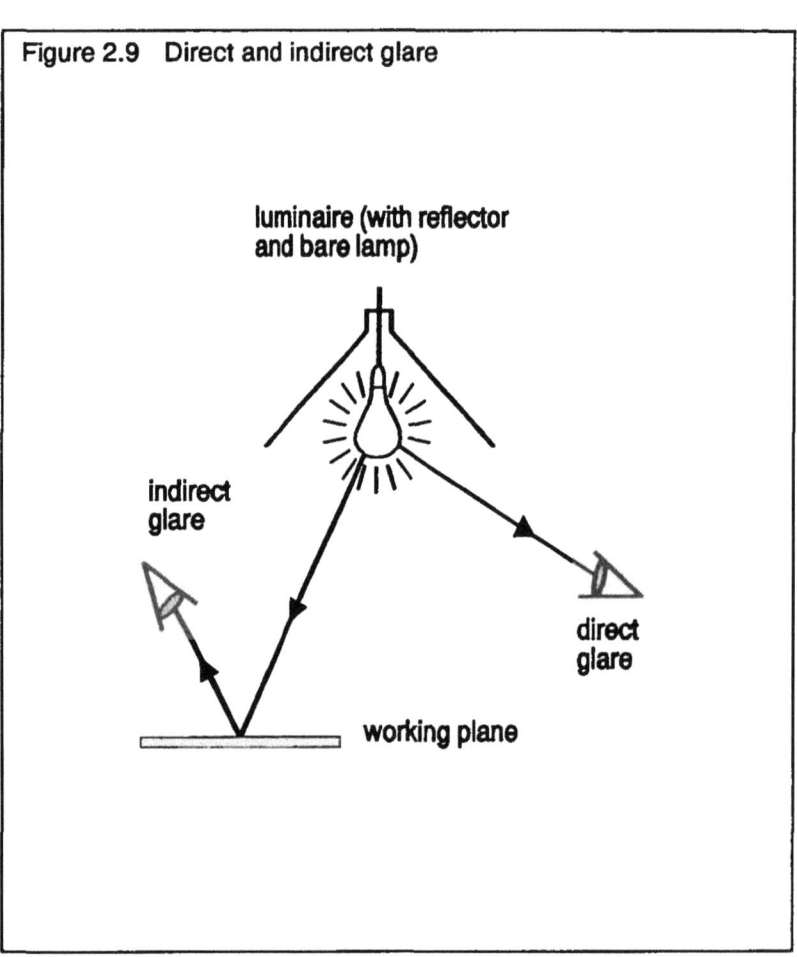

Figure 2.9 Direct and indirect glare

luminaire (with reflector and bare lamp)

indirect glare

direct glare

working plane

2.8 Illuminance For Different Purposes

In urban areas where energy is less of a problem, the decision may be based on the CIE recommendations for the minimum levels of illuminance for various interiors and visual tasks. Table 2.5 gives the scale of illuminances recommended by the CIE for use in working interiors. This represents the most basic international consensus about lighting. The standard illuminances are based on consideration of the performance of the respective tasks, the comfort of people doing the tasks, and the time for which the space is occupied.

Table 2.5 indicates, for each type of activity, a range of illuminances which would be appropriate, but each range covers two to three discreet steps of the scale with a ratio of about 2:1 or more in illuminance. Where energy is widely available, professional lighting designers should generally apply their country's national standards. These can vary

considerably from country to country, and it is impossible to reproduce all of them in this book.

The left-hand column of Table 2.6 gives examples of a few design service illuminances recommended in the United Kingdom for specific interiors and activities (these are based on the Chartered Institution of Building Services Engineers (CIBSE) Code for Interior Lighting).

Levels of illuminance attained in practice

Inside a building not devoted to accurate visual work anything over 500 lx is by today's standards considered fairly generous. In most domestic living rooms however, the accepted illuminance is **about 50 lx**. The human eye, then, is very versatile. Indoors, the natural

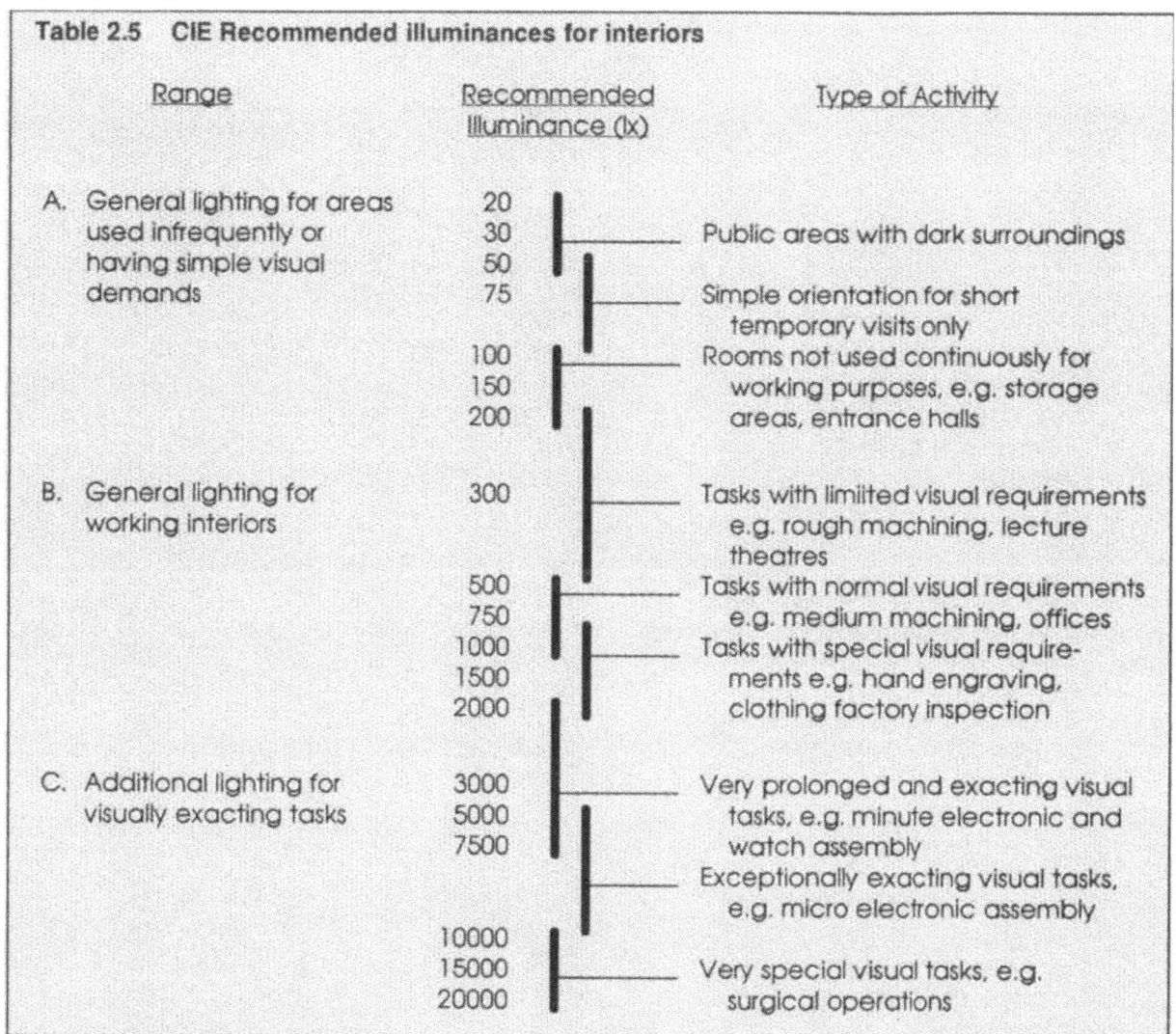

Table 2.5 CIE Recommended illuminances for interiors

Range	Recommended Illuminance (lx)	Type of Activity
A. General lighting for areas used infrequently or having simple visual demands	20 30 50 75 100 150 200	Public areas with dark surroundings Simple orientation for short temporary visits only Rooms not used continuously for working purposes, e.g. storage areas, entrance halls
B. General lighting for working interiors	300 500 750 1000 1500 2000	Tasks with limited visual requirements e.g. rough machining, lecture theatres Tasks with normal visual requirements e.g. medium machining, offices Tasks with special visual requirements e.g. hand engraving, clothing factory inspection
C. Additional lighting for visually exacting tasks	3000 5000 7500 10000 15000 20000	Very prolonged and exacting visual tasks, e.g. minute electronic and watch assembly Exceptionally exacting visual tasks, e.g. micro electronic assembly Very special visual tasks, e.g. surgical operations

tendency is to take something to the window in order to examine it in detail. Humans, after all, are evolutionary animals adapted to exterior light, and human vision under other circumstances is comparatively inferior. Where people are given a free choice, they will spontaneously select levels of 1000 to 2000 lx! It is therefore not surprising that buildings such as supermarkets or shopping malls are often lit to this level of illuminance.

Minimum acceptable illuminance levels

In a perfect world, illuminances similar to those above would be the norm. However, light from natural (e.g. windows) or artificial sources represents costs both in terms of money and energy. In the industrialized world, economic progress has led to a situa-

tion in which there is little distinction between what people need and what people want. In lighting terms, this leads to higher than necessary levels of illuminance in many places.

In the rural areas of developing countries, what people need and can afford are usually the relevant criteria. Hence, it may be appropriate to adopt the concept of 'pre-electrification'. Pre-electrification consists of providing a community with a minimal quantity of electricity which is enough for a few low-power but efficient lamps. For example, this small quantity of electricity can be provided by a small generator set, a solar photovoltaic installation or a small wind turbine generator. In the right hand column of Table 2.6 the level of illuminance suggested constitutes a great improvement over the use of kerosene lamps although remaining far behind recommended

Table 2.6   Choice of design service illuminance

Location	Level when grid available lx	Level minimum pre-electrification lx
EDUCATION / BUSINESS		
Office desk	300 / 500	50 / 100
Office ambience	300 / 500	15
Reading area	300 / 500	25 / 50
Classroom	300	25 / 50
Corridor	100 / 500	5
Workshop ambience	> 200	10 / 50
Work spotlight	> 300	20 / 100
HEALTH CARE		
Examination room spotlight	> 500	500
Delivery room spotlight	> 2000	1500
Delivery room ambience	> 200	50
Post delivery room	150	15
Surgery room spotlight	> 2000	1500
Surgery room ambience	> 200	50
Rest room	150	15
DOMESTIC LIGHTING		
Dining room/ restaurant	50 / 200	25
Livingroom	50 / 300	15
Bedroom	50	15
Kitchen working area	300	25
Bathroom sink light	> 300	50 / 100
Toilet	20 / 100	10
OUTDOOR LIGHTING		
Street / outdoor areas	5 / 20	2.5
Sign illumination	300	200

CIE illuminances when a reliable mains electricity is available.

The right-hand column values in Table 2.6 may seem very low compared to those of the left column. However they are high enough to ensure enough light, although not with all the comfort of the standard illuminances. If the tasks are to be performed for unusually long periods or if the users have poor eyesight (e.g. elderly people), the level of illuminance should be increased. In a few cases, it may be possible that the system would need upgrading slightly soon after installation. This is a more efficient way to design a system than using standard recommended high illuminances that will be unaffordable both in capital and running costs.

Contrast in lighting

It may not be sufficient to achieve just the required design illuminance on the working plane. Illuminances on other planes (e.g. the walls and floor) should ideally also be kept in proportion in order to avoid visual discomfort. According to some standards, in an interior with **general lighting** (i.e. lighting designed to illuminate a whole area uniformly, without provision for special requirements) the ratio of the minimum illuminance to the average illuminance over the task area should not be less than 0.8.

In an interior with **localized lighting** (i.e. lighting designed to provide the required service illuminance on work areas, together with a lower illuminance for other areas) the ratio of the illuminance on the task area to the illuminance around the task area should not be more than 3:1. For **task lighting** (i.e. lighting designed to provide illumination only over a small area) the ratio should also not be more than 3:1. Chapter 5 describes in some detail the various lighting design classes: general, localized and local lighting.

However, in most pre-electrification cases, it is not advisable to follow these rules strictly because they usually require much more light and therefore more energy which in turn is often unaffordable for the rural population.

2.9 Measurements in Lighting

Strangely, the eye cannot be relied upon to tell how much light is present. The eyes' built-in automatic adjustment mechanism varies the size of the pupils and their reactions to colours and contrasts will usually combine to give an answer that is unlikely to be accurate. It is usually possible to tell whether the lighting level has been increased or decreased, but not by how much.

Light is generally measured by means of a luxmeter. This consists of a photo-electric cell which transforms the incident radiation into electricity. Luxmeters are compact, easily transported and measure illuminance in 'lux'. Usually a simple pocket type costing around US$150 is adequate and accurate enough (e.g. +/- 10%).

Figure 2.10 shows a simple luxmeter. More accurate and more expensive instruments are available if needed. In all cases, luxmeters need to be well maintained to keep their accuracy and should be checked against others known to be correct at regular intervals.

The photo-electric cell of the luxmeter will not respond in exactly the same way as the human eye to different coloured light, and therefore a correction factor (obtained from the manufacturer or included with the luxmeter) should be applied when the light source used is different to the kind (usually incandescent lamps or lamps with similar colour rendering) under which the instrument has been calibrated. In practice this may also be achieved by means of fitting the luxmeter with special filters.

Measurement techniques

The photocell must be parallel with the surface on which the illuminance is to be measured. For example, to measure the illuminance on a reading table, it must face directly upwards. The measurements should be conducted in real working situations. What concerns a working person is the illuminance in his or her normal working position. If, when he or she is there, the illuminance is only half that when he or she is not there, the location of the luminaire may need to be changed.

When taking readings of illuminance from discharge lamps (e.g. fluorescent tubes and

Figure 2.10 Pocket luxmeter, with photocell (right)

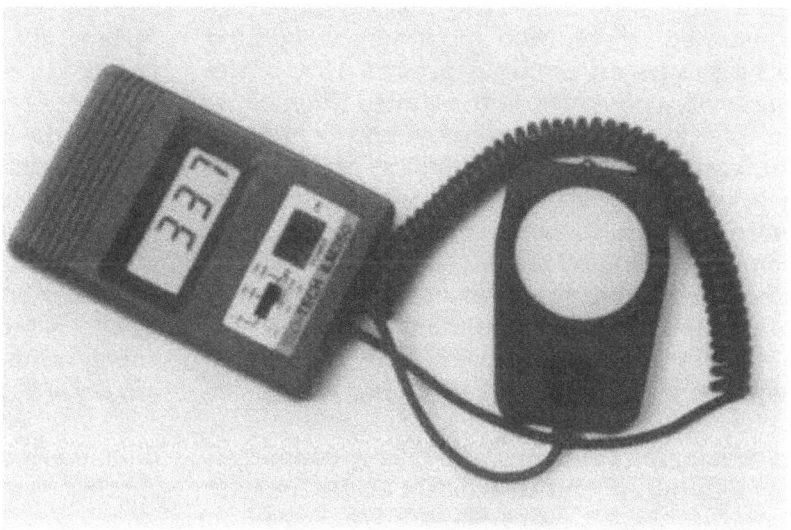

CFLs), allow at least ten minutes after turning the lamp on in order to allow the illuminance to stabilize.

For measurements 'in situ' one cannot expect a precision greater than +/- 10%. This should be kept in mind when assessing the effectiveness of a lighting installation. In order to determine accurately the average illuminance in a room, the whole working plane should be divided into a number of equal areas (e.g. one square metre) and the measurement carried out in the centre of each area. At least 20 measurements are necessary for an accurate result. Ideally, lamp operating temperature and mains voltage should be checked for consistency before and after carrying out light measurements.

When assessing a lighting installation, it should be remembered that a good lighting system is one which makes the users' activities easy to perform, which provides them with a sensation of comfort, and which complies with aesthetic requirements. All these elements tend to be subjective and are difficult to quantify.

2.10 Energy Efficiency: Towards Electrical Lighting

On a worldwide basis, the quantity of energy consumed for domestic and industrial lighting is huge . As an example, around 15% of the electricity consumed in the United Kingdom is for lighting, costing industry and commerce the equivalent of US$1.8 billion per year. In developing countries, accurate figures are more difficult to obtain, but the energy for domestic and industrial lighting still represents a significant fraction of the energy consumed. As the population increases and hopes for better living conditions, the energy required for lighting is likely to increase.

In developing countries, the energy consumed for lighting in rural households is not as

Figure 2.11 A kerosene lamp (left) and a solar portable lantern with a 4W fluorescent tube (right): 'Which lamp gives twice as much light and consumes no fossil fuels?'

much as the energy consumed for cooking. This may explain why many development programmes have so far concentrated only on improving the energy efficiency of cookstoves. Usually, improved cookstoves increase the efficiency of burning biomass fuels from an initial 10% to reach 30% at most. For lighting systems, the scope for energy efficiency improvement is much greater.

For example, by replacing kerosene lamps, which are used by the million today, with a 4W electrical fluorescent lamp, the energy-to-light efficiency of the lighting system is multiplied by a factor of 100. Moreover, the light output is increased by a factor of at least two. If the electrical fluorescent tube is powered by a small photovoltaic module or a small wind turbine, the primary energy use for lighting is reduced to zero.

Increasing energy costs for users combined with increasing environmental costs to society as a whole (e.g. global warming due to carbon dioxide from increased power generation) place the lighting designer or specifier in a position where he or she should ideally seek a design which will achieve the maximum energy conservation, without reducing the quality of lighting.

Whenever possible, it is advisable to choose a lighting system with the highest luminous efficacy, and consequently with the highest energy efficiency.

Great care is needed to design each lighting system component to achieve the lowest energy consumption for a given light need. If not, only a fraction of the light output of the lamp will be useful on the working plane, either because the luminaires (i.e. fittings, reflector and diffuser) and the walls will absorb too much of the light output of a lamp or because too much light will reach unwanted areas.

Chapters 5 and 9 give advice on how to choose and design an efficient lighting system considering all the system components and the light requirements.

2.11 Units Used in Lighting Designs

The SI (Systeme International) of metric units has become the most widely adopted international standard for most forms of measurement, including lighting. Hence in this book, the photometric and energy quantities are expressed in SI units. However, non-SI units (e.g. foot-candle to measure illuminance or the BTU to measure energy) are still being used in some countries. To help the reader to convert non-SI units into SI units and vice versa, a conversion table is given in Appendix F. The table covers the more common photometric units that are used in lighting design, as well as units of energy, power, length, capacity and temperature.

Flame-based Lighting

3

3.1 Background

The history of flame-based lighting started when humans learned to control fire. This was used to protect themselves from animals and cold weather as well as for artificial lighting. Since that time flame-based lighting systems have been designed and improved, starting from wood torches to the fairly sophisticated kerosene pressure lamp. While in the industrialised world flame-based lighting systems have been almost totally replaced by electrical lighting systems, millions of flame-based lighting systems are still being used in developing countries, where either there is no access to the electricity grid or where people cannot afford electricity.

The production of light by burning carbon-based fuels such as wood, vegetable oil, kerosene or gas is based on the principle of incandescence. This occurs when a solid substance is heated to such a high temperature that it emits visible radiation (i.e. light). The radiation emitted depends not only on the temperature of the substance but also on the spectral emissivities (i.e. some substances emit more visible radiation than others at a given temperature).

When burning fuels, small particles of carbon in most luminous flames give a white light made up from a broad band of wavelengths in the visible spectrum. This gives a 'natural' colour and excellent colour rendering. The lamps using this principle (e.g. candles and kerosene lamps) can be classified as carbon incandescence sources.

However, when burning gas or vaporised liquid fuels, the visible light output can be increased by putting a rare-earth incandescent mantle over the flame. Since the invention of incandescent mantles at the end of the 19th century, all gas lamps and oil vapour lamps (e.g. kerosene pressure lamps) use rare-earth incandescence to produce light. They are classified as rare-earth incandescent sources.

The present day

Today the main flame-based light sources are of two principal types:

- Incandescent sources
  - Open fire
  - Candles
  - Oil lamps
  - Kerosene lamps
  - Carbide lamps

- Rare-earth incandescent sources
  - Kerosene pressure lamp
  - LPG (Propane/butane) lamps
  - Methane/biogas lamp

Figure 3.1   A kerosene pressure lamp in India

18

Each of these sources will be examined in the following sections. The last section summarizes the main characteristics of the different light sources.

3.2 Carbon Incandescent Light Sources

General principle

Light is produced by burning carbon-based fuels such as wood, vegetable oils or kerosene. The production of light from these sources is based on the principle of carbon incandescence, which occurs when carbon is heated to such a high temperature that it emits visible radiation.

The small particles of carbon in most luminous flames yield a white light made up from a broad band of wavelengths in the visible spectrum. This has a 'natural' colour and an excellent colour rendering. For example, the white-yellow flame of a candle is produced by 'white-hot' particles of carbon (i.e. soot) in the flame.

Traditional open fire

Since humanity learned to control fire, it has been actively used as an aid in hunting, clearing land for agriculture, cooking food obtained from these activities and, last but not least, lighting the cave or the hut.

Fire-baskets or 'cressets' standing on high poles and fed with very dry wood were used in the past for lighting in Egypt, Arabia, India and Europe. Even today in many developing countries, particularly sub-Saharan Africa, wood remains the most common, if not the only, fuel used in the rural areas for cooking. An open wood fire will provide an illuminance of about 1 to 5 lux at a 2m distance. This is sometimes the only source of light for rural people and the need for light is often a factor against the use of more effi-

cient and enclosed cooking stoves. The failure to recognize the importance of open-fire lighting has caused difficulties for many recent improved wood-stove dissemination programmes.

The need to manage wood resources is usually seen as very important by the managers of environmental or energy-related projects, but not always by villagers who do not understand the link between removing the tree cover and the decline in rainfall. As an illustration, a forestry project manager in Mali reports that in some villages, tens of wood fires are lit along the lanes to provide orientation lighting for people gathering during night-time festivities. This and other examples show that the need for improved lighting systems is at least as important as the need for improved cook-stoves. It would therefore make sense for each improved cooking stove programme to include a lighting sub-programme and vice versa.

Candles and torches

Operating principles and construction

Burning brands of wood taken from a fire and carried into the surrounding darkness were the most primitive form of torches. Then came the use of thin splinters obtained by splitting straight logs of knot-free resinous wood (such as pine). Dried rush stems were also used as torches. These so-called 'rush-lights' were the earliest form of 'candles', made by stripping

Figure 3.2 Traditional open-fire 'three stones' in Mali

Candles: quick data	
Colour rendering:	Good
Luminous efficacy:	Very poor (0.02 to 0.2 lm/W)
Luminous flux:	1 to 16 lm
Fuel consumption:	5.5 to 7.2g of wax/h
Life:	3h to 5h (for a 2cm-diameter and 12cm-high candle)
Energy source:	Paraffin wax (self-contained)
Applications:	General, localized and local lighting for homes.
Comment:	Very small light output that can be enhanced by a reflector. Produces black smoke (soot). Fire hazard. Very easy to use.

rush stems of all but the thinnest rinds. The rinds were dried and then repeatedly dipped into hot animal fat and cooled until a thick wall of fat was built up. Such 'candles' were extensively used by the Egyptians several thousand years ago.

The first true candles were made by the Romans, who used wicks made from a species of papyrus, with tallow or wax as fuel. Wax was obtained from bees' honeycomb. In Europe, from the Middle Ages until the nineteenth century, candles were mostly made of tallow obtained from mutton, pork or beef fat, which people carefully saved from their kitchen wastes. Tallow candles were home-made products.

By the end of the eighteenth century a process was discovered for hardening tallow, and stearine and other fatty acids were introduced into candle manufacture. Animal horn was processed into thin translucent glazing for early lanterns, to protect the flame from being blown out by the wind. Ordinary candles are currently manufactured from paraffin wax which is a by-product of the petroleum refining industry.

Performance and applications

The light output of currently manufactured ordinary candles ranges between 1 and 16 lumens. The latter is only one twentieth of the light output of an 8W fluorescent tube. The candles tend to burn quite quickly (5 to 7 grams per hour for a 2cm diameter candle). This means that an ordinary household candle of 2cm diameter will shorten at a rate of about 3cm/hour. By protecting the candle with a glass cylinder or even a paper sleeve, the consumption rate is reduced to 2 to 3 grams per hour, but as a result the light output is also reduced to approximately 2 lumens.

The light produced by a candle is emitted in all directions. Unless the candle is in the centre of a room or a table, in which case all the light can be used, much light is absorbed by the nearby environment (e.g. walls): consequently the useful light is drastically reduced. To overcome this problem a candle situated near a wall can be backed by a reflector (e.g. plane or circular) which can be made of aluminium or an unfolded tin can.

The light output of a candle positioned in a room's corner can be enhanced in the same way if the candle is fitted into a 'luminaire' with reflectors on both sides of the adjacent walls. This gives good results by reflecting most of the candle light into the room. Furthermore it gives the agreeable visual impression of seeing several candles while only one is burning!

In the developed world candles are now used only for religious meetings, ceremonies, decoration and in some cases, emergency lighting in households. In developing countries, candles have the same uses but they are also used for regular domestic purposes and in health centres and schools. Prices of candles vary a lot according to size, quality, etc. As an indication, the price of a candle of 2cm diameter and 12cm long is in the range of US$0.10 to $0.15.

Oil lamps

Operating principles and construction

The story of the oil lamp began in the prehistoric era when humans discovered that a wick of vegetable origin, if soaked with animal fat, produced a lasting light while burning. Hollow stones were used as oil lamps as long as 15,000 years ago. Since then, several materials have been used or made into lamp containers: seashells, pottery, glass, bronze and iron.

In early lamps, the wick would hang over the edge. Later, channels, spouts and floats were used to support the wick. Wicks were made from any available fibrous material which would, by capillary action, draw the fuel from the container and feed it to the wick-hole. Depending on the region, vegetable oils (olive, sesame, nut, etc.), and animal oils such as fish or whale oils were used as fuel. Since these oils were edible, they were used for cooking as well.

It is perhaps surprising to note that from the Stone Age up to the nineteenth century, while lamp design evolved greatly aesthetically, little

Figure 3.3   Open saucer and wick spout oil lamp

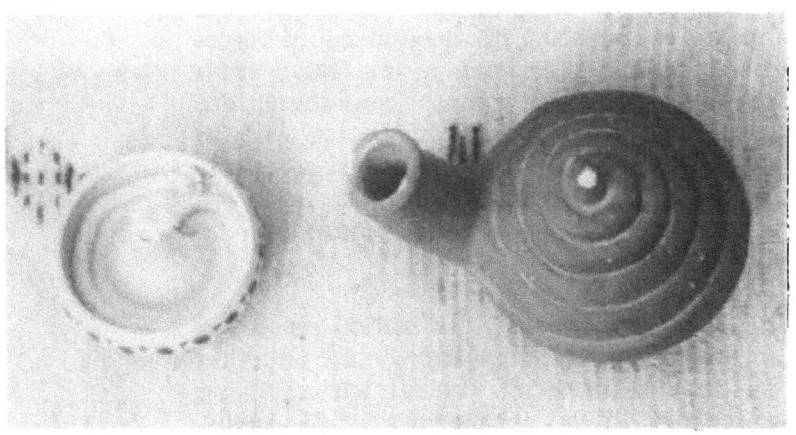

progress was achieved in improving the performance of oil lamps. The only major improvement was introduced by Argand in 1782-4: by introducing a chimney and a tubular wick, air would circulate in and around the flame, so that the combustion was complete and smokeless, giving a much brighter light. Unfortunately, the Argand burner consumed so much oil (about 12 times more oil than traditional designs) that it was only used by rich people. Traditional oil lamps continued to be in use until quite recently even in Europe: in London, street lighting by oil lamps was first introduced around 1800.

Simple oil lamps are still in use by very low-income households in many developing countries. In India and many African countries,

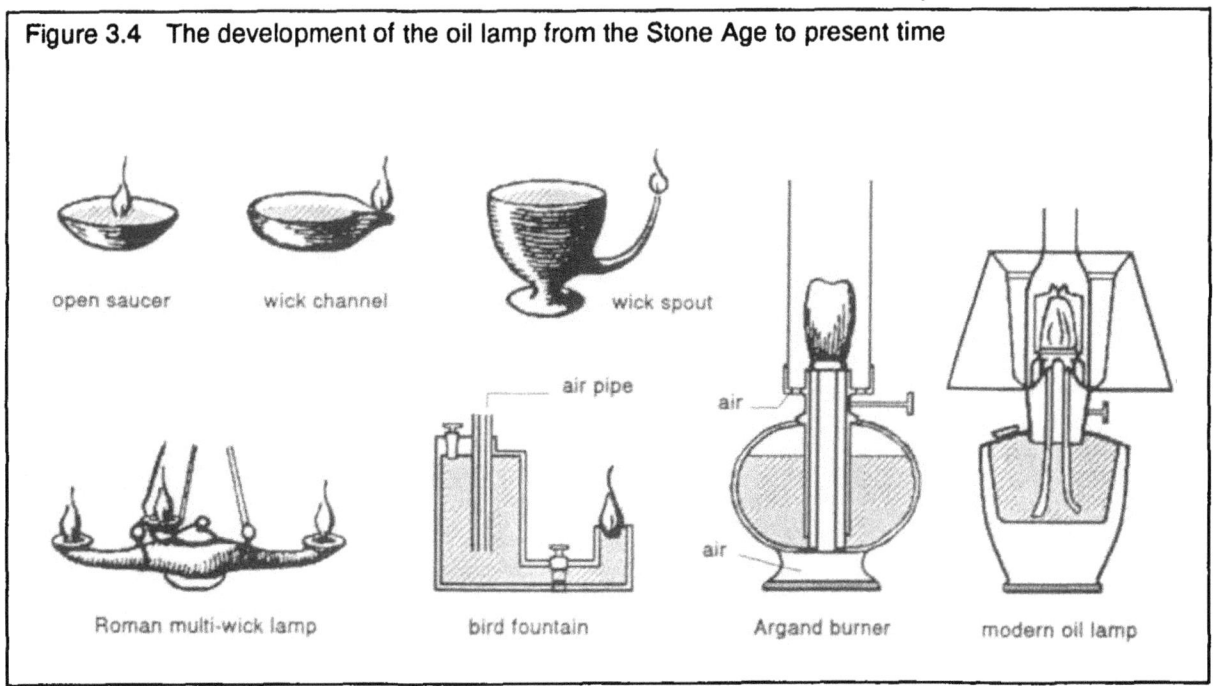

Figure 3.4   The development of the oil lamp from the Stone Age to present time

open saucer   wick channel   wick spout

air pipe

air

Roman multi-wick lamp   bird fountain   Argand burner   modern oil lamp

traditional oil resources such as shea butter or palm oil are used as fuel. Traditional oils and lamps are being replaced in many places by kerosene. Tin can lamps (e.g. made from small concentrated tomato containers) are quite widely used by the poor people living in slums and by hawkers selling food alongside the roads. Sometimes, the flame is protected against the wind by a plastic bottle with its ends cut off.

The discovery of cheap mineral oil resources in 1859 in Pennsylvania and the introduction of such features as central wicks, flame aeration and glass chimneys were the beginning of a rapid series of progressions, which led to the development of more modern lighting appliances. Figure 3.4 illustrates the development of the oil lamp from earliest times to the present day.

Performance and applications

It is very difficult to define the performances and life of oil lamps because they exist in so many shapes and sizes. However, a tin-can lamp can be assumed to have a similar light output to a candle, i.e. from 1 to 16 lm. A large oil lamp with a glass-covered mantle may have a light output similar to a kerosene lamp, i.e. 10 to 100 lm.

Kerosene lamps

Kerosene lamps (also known as hurricane lanterns) are presently the most common source of light in developing countries for rural households. They are also a type of oil lamp which is why they deserve a special

Figure 3.5 Traditional oil lamps still being used in Mali. In the background, an old kerosene lamp

section here. In India alone, the number of kerosene lamps in use exceeds 100 million. Hurricane lanterns were first introduced in the 1880's to burn mineral oil, usually kerosene, safely, economically and with minimal attention.

Operating principles and construction

In a kerosene lamp, light comes from the glow of a luminous yellow flame produced by a cotton wick impregnated with kerosene. The kerosene reservoir forms the base of the lamp and is closed with a cap usually attached by a small chain to the lamp. The flame is prevented from being blown out by a protective glass bulb. This needs frequent cleaning because

Kerosene lamps: quick data	
Colour rendering:	Good
Luminous efficacy:	Very poor (0.05 to 0.2 lm/W)
Luminous flux:	10 to 100 lm
Fuel consumption:	0.02 to 0.05 l of kerosene/h
Life:	2 to 4 years, but requires spare parts (wicks, glass-covered mantle)
Energy source:	Kerosene
Applications:	General, localized and local lighting for homes and outdoors.
Comment:	Small light output, portable, emits soot which decreases the light output and blackens the surrounding walls. Possible fire hazard.

soot accumulates on it from the flame. The light output is considerably decreased by a dirty bulb. The overall design of the lamp allows it to withstand winds up to 10 metres per second.

The light output of the lamp can be increased or decreased by adjusting the height of the burning wick by turning the adjuster knob. There are many sizes of hurricane lamp, but the light output is ultimately dependent on the width of the wick. The wider the wick, the greater the light output will be and vice versa.

For a large kerosene lamp, the standard large wick can be replaced by a smaller one for lesser lighting needs. The production of soot can be minimized by cutting the wick horizontally and cleanly, leaving no fraying strands. Unfortunately, the burning wick does not always keep this shape and therefore frequent trimming may be necessary.

Performance and applications

The light output depends upon many factors such as size of the lamp, width of the wick and length of the burning wick. However, on average the light output is low, from approximately 10 lm at low power to 100 lm at full power. By comparison, 100 lm is one third of the light output of an 8W electrical fluorescent lamp or one tenth of that of a 100W incandescent electrical lamp. The kerosene reservoir is usually large enough for the lamp to burn for 7 to 10 hours at full power.

Kerosene lamps are relatively inexpensive, simple to light, portable, and consume very little kerosene. However, they require frequent cleaning and maintenance, and they are not efficient as light producers. Glass breakage is also a common problem and can be relatively expensive for some users to replace. Users must therefore buy wicks and glass bulbs as well as a supply of kerosene on a regular basis. The price range for these lamps is from US$5 to $10 depending on the size and quality (i.e. durability) of the lamp.

Figure 3.6   Two sizes of kerosene lamp

As mentioned earlier, kerosene lanterns are used widely where there is no electricity grid. Even in areas where grid electricity is available, it is common to find households who cannot afford electricity still using hurricane lamps. Also, since the electricity supply is often unreliable in many countries, households are often equipped with kerosene lamps for use during power cuts.

Carbide lamps

Carbide or acetylene lamps are rather uncommon today. However, they were widely used in Europe in the 1890s, when electricity was expensive; they were used for headlights for cars and for bicycles on account of their bright flame that can be focused into a beam with a mirror and which is less easily extinguished by draughts than an oil-lamp flame.

Operating principles and construction

A carbide lamp burns acetylene which is produced in the lamp from a chemical reaction between water and calcium carbide. The acetylene produced can be lit with a flame and burns in air with a bright luminous flame well suited as a light source.

Acetylene is obtained from the chemical combination of water with calcium carbide according to the following reaction:

23

Acetylene lamps: quick data	
Colour rendering:	Good
Luminous efficacy:	Poor (1.46 to 1.90 lm/W)
Luminous flux:	50 to 250 lm
Fuel consumption:	6 to 23g of carbide/h
Life:	3 to 5 years
Energy source:	Carbide
Applications:	General, localized and local lighting for homes and outdoor use.
Comment:	Adjustable light output, portable, light output depends on cleanliness of the reflector.

$$CaC_2 + 2H_2O > Ca(OH)_2 + C_2H_2$$

calcium carbide + water > slaked lime + acetylene

Approximately 1kg of calcium carbide and 0.50kg of water should produce 1.15kg of slaked lime and 0.40kg (or 340 litres) of acetylene. In practice only 300 litres are obtained from commercially available calcium carbide, because it contains 20% calcium oxide along with other impurities.

In most acetylene lamps a water container is placed above the carbide container so that the water is fed onto the carbide by gravity. Gas pressure in the carbide container automatically controls the speed at which water enters the container; if too much water enters, more gas is generated and the pressure rises, reducing the flow of water. There are some designs, in which calcium carbide is contained in a small basket, which is more or less immersed into water to produce the required quantity of gas. Figure 3.7 shows a typical commercial acetylene lamp.

The use of a carbide lamp is fairly simple. It is first filled with a carbide charge and with water. The water tap is then opened which allows water to drip onto the calcium carbide. Once the acetylene is produced, the lamp is lit by a built-in ignition flint or with a match. The lamp is turned off by closing the water tap. The light output can be adjusted in a range of 30 to 100% by turning the water tap. Once the lamp is switched off, it will produce acetylene for a little while giving a typical 'garlic-like' smell. The lamp can be switched on again until the carbide charge is fully consumed. The resulting slaked lime is an alkaline substance that should be disposed of with care.

Performance and applications

Acetylene produces a relatively concentrated bright white light which can be focused much more readily than a kerosene wick flame. This feature makes acetylene lamps attractive for many professional applications. The light output of acetylene lamps is also much greater than that of kerosene wick lamps, varying from 50 to 250 lm, and their fuel consumption rate is from 6 to 23g of carbide per hour. Typically, a carbide lamp with a charge of 50 to 70g of calcium carbide will run for 4 to 5 hours. As an indication, a carbide lamp with a 150mm reflector, giving a light output three times that of a wick kerosene lamp, may cost US$35 to $50.

In developed countries, acetylene lamps have been replaced by electric lamps since the

Figure 3.7   Carbide lamps (courtesy of Caving Supplies Ltd)

beginning of the century, but there is still a small leisure market for them, mainly for caving and pot-holding activities.

In developing countries interest in the use of acetylene lamps for rural lighting rose again at the beginning of the 1980s with the improved domestic stoves programmes, where fire light had been largely excluded to obtain higher cooking efficiency, but little has been done to evaluate and popularize them. Although carbide-based lighting is rarely used anymore, it may deserve more serious consideration for lighting in non-electrified areas.

3.3 Rare-earth Incandescent Lamps

Basic principles and history

Early attempts to use coal gas for lighting were made by Murdoch in 1792, when he illuminated his house with coal gas. Independently, Winsor demonstrated the first gas street lights in London (Pall Mall) in 1807. These early lights used simple gas burners, which were merely holes in gas pipes or apertures left when the ends of pipes were partly blocked. The Argand burner was later adapted for gas and widely used. All these early gas lights used the incandescence of carbon particles contained in the gas to produce light and therefore had a very poor energy-to-light conversion efficiency.

R.W. Bunsen in 1855 improved the Argand burner in order to obtain a very hot but almost non-luminous smokeless flame while C.A. von Welsbach discovered in 1880 the possibility of making incandescent mantles using uncommon chemical substances such as thorium oxide instead of the metals which had been used in earlier experiments. Finally in 1893, he used a cotton mantle impregnated with a mixture of rare earths (cerium and thorium) fitted onto a Bunsen burner, which glowed brightly in the flame and gave much more visible light than the flame alone.

The introduction of the rare-earth incandescent mantle was a great improvement in gas lighting which led to the development of the rare-earth incandescence lamps which have an energy-to-light conversion efficiency 10 times greater than the early gas lights.

The production of light from these sources is based on the principle of rare-earth incandescence, which occurs when rare-earth compounds are heated to such a high temperature that they emit visible radiation. In contrast to carbon, the spectral emissivities of the rare earths are selective; i.e. when heated, rare-earth compounds emit most strongly at

---

Caution with rare-earth incandescent mantles

Today, most incandescent mantles consist of 90% radioactive thorium oxide (thorium-232) and the rest is made of cerium oxide which makes the light appear whiter. However, when burning, the mantle emits radioactive particles into the atmosphere and the shock-sensitive mantle can easily crumble to fine radioactive dust which may by ingested by the user. It is well known that any radiation dose adding to natural radioactivity may lead eventually to health hazards. To overcome the problem, some manufacturers (e.g. Coleman) have launched a thorium-free mantle where thorium has been substituted by ytrium. However, most of the mantles in use today still contain thorium. Hence, it is recommended to:

- Store no more incandescent mantles than are necessary;
- Store incandescent mantles out of reach of children;
- Buy thorium-free mantles or those with the lowest possible radioactivity;
- Use lamps that offer little access to insects and wind;
- Do not install lamps above places where food is stored or processed;
- Remove old incandescent mantles with care (e.g. collect them and their ashes in a plastic bag) and wash your hands afterwards.

certain wavelengths. As a consequence, the spectral composition of the light from an incandescent mantle differs from that of incandescent carbon, or daylight, and its colour is therefore 'unnatural', so colour rendering is poor. However, the reason for using rare-earth incandescent mantles over a gas flame is to obtain far more light than would otherwise be possible. To do this, a hot and relatively non-luminous gas flame is used to heat an incandescent mantle to the high temperature needed to make it radiate. The use of liquid fuels, such as kerosene, gasoline or alcohol, requires special construction to provide for pre-heating of the fuel in order to vaporize it into a gas prior to combustion. A hot non-luminous flame is essential to burn up all the carbon in the fuel and thereby avoid deposition of carbon onto the incandescent mantle, which would otherwise eventually damage it.

Today all gas and oil vapour lamps are fitted with Welsbach's (i.e. rare-earth) incandescent mantles, and produce a light output which compares quite favourably with 60 to 100W electric standard incandescent lamps.

Kerosene pressure lamps

This type of kerosene lantern is frequently called a 'Tilley Lamp' in English-speaking

Figure 3.8   Lighting a kerosene pressure lamp in a bar, Zaire

areas and a 'Petromax' in French-speaking countries.

Operating principles and construction

This type of kerosene lantern is a pressurized lantern, where a small handpump is used to pressurize the air trapped above the fuel in the fuel tank. This forces a steady supply of kerosene through a pre-heated fuel tube in order to obtain kerosene vapour, sometimes called kerogas. The kerogas is ejected under pressure from a fine nozzle and mixes with air near it. The fuel-air mixture is then burnt to produce a high-temperature, non-luminous flame, which heats the mantle (and also heats the fuel tube for further kerogas formation).

Methanol (methyl alcohol or wood spirit) is commonly used to pre-heat the tube suffi-

Kerosene pressure lamps : quick data	
Colour rendering:	Poor
Luminous efficacy:	Poor (0.4 to 1.6 lm/W)
Luminous flux:	220 to 1300 lm
Fuel consumption:	0.06 to 0.08 l of kerosene/h
Life:	5 years, but requires spare parts
Energy source:	Kerosene
Applications:	General, localized and local lighting for homes and outdoor use.
Comment:	Noisy, high consumption of kerosene, portable, explosion risk, not always easy to use. There is also a health risk from mantles.

ciently for the kerosene to vaporise when starting such a lamp. Some modern kerosene pressure lamps can be pre-heated with only kerosene making the lamp more convenient to light.

Performance and applications

Pressure lamps are portable and robust and they produce much more light than hurricane lamps. According to various sources, the light output of a kerosene pressure lantern ranges from 200 to 1300 lumens, which may be compared with the light output of a 100W electric incandescent light bulb. On the other hand, they are expensive (e.g. US$50 to $60),

Figure 3.9 Gas lamp powered by a 13kg butane container in Syria

heavy, noisy, smelly and difficult to light. They require frequent pumping, and consume a lot of fuel 0.06 to 0.08 litres of kerosene per hour. They also emit a lot of heat which makes them unpleasant to use in hot climates.

New designs are being developed in order to increase the efficiency of pressure lamps and reduce operating costs. One of these designs, called the 'Noorie', is being developed by researchers in India. It works on the same principle as traditional designs of kerosene pressure lamp. and consumes 0.05 l of kerosene per hour while producing 1250 lm. Hence, it produces a light output equivalent to that of a standard kerosene pressure lamp, but with only 60% of the kerosene consumption and about one-third of the pressure used in them.

The Noorie lantern can also run on ethanol (with a consumption rate of 65g/h and a light output of 1270 lm). Ethanol, which can be derived from biomass, can be a renewable replacement for kerosene as a lighting fuel. The Noorie lantern can also be used as a cooking device because of the inherent heat output which is typically around 500W. This is similar to the heating power of a small cookstove!

Because of their high capital and running costs, kerosene pressure lamps are not used by poor people but rather by bar keepers, rich merchants and relatively wealthy traders who can profit by attracting clients with good lighting on their premises.

Propane/butane lamps: quick data	
Colour rendering:	Poor
Luminous efficacy:	Poor (0.9 to 2.4 lm/W)
Luminous flux:	330 to 1000 lm
Fuel consumption:	28 to 34g of gas /h
Life:	2 to 5 years, but requires spare parts (e.g. mantle)
Energy source:	Butane/propane
Applications:	General and localized lighting for homes and outdoor use.
Comment:	Quiet, very convenient to use, portable, possible explosion hazard, health hazard from some mantles.

LPG (Propane/butane) lamps

These lamps have been largely marketed in developed and in developing countries by large companies such as Camping Gas International. They are often powered by a small gas container (often of blue colour) holding the LPG (liquefied petroleum gas).

Operating principles and construction

Gas lamps use a similar type of rare-earth incandescent mantle to the kerosene pressure lamp because natural gas (i.e. methane/propane/butane) burns with a blue, non-luminous flame; they therefore require the use of an incandescent mantle to produce light reasonably efficiently. Some gas lamps consist only of a small burner with a rare-earth mantle around it. These are cheap but cannot work outside in the rain. They can be easily blown out by the wind and be rapidly soiled or even destroyed by dust and insects. More expensive models have a glass 'chimney' around the incandescent mantle which acts both as protection and as a light diffuser.

Gas lamps are superior to kerosene pressure lamps in the sense that they produce a comparable quality and intensity of light but there is no need for pumping them up; also the gas ignites immediately, obviating the need for pre-heating by burning alcohol. Lighting a butane or propane light is as simple as turning on the gas tap and lighting the gas with a match.

Users should always verify that when the lamp is not in use, the gas tap is firmly closed to prevent explosion caused by gas leakages. The most widely-used gases for lighting are LPG (liquefied petroleum gas) which can be propane or butane. Such gases are available in re-fillable or disposable cylinders (see Chapter 6).

Performance and applications

Gas lamps give the same light output as kerosene pressure lamps, typically 330 to 1000 lm. The latter figure is equivalent to the light output of a 75W incandescent electric bulb. Gas consumption is typically between 28 to 34g per hour. The light output can be adjusted using the gas tap.

Gas lamps also tend to produce less odour than kerosene lamps (which can smell quite unpleasant). The purchase cost of the lamp, excluding the gas container, is in the range US$10 to $35, depending on whether the lamp has a glass mantle cover.

Gas is a very convenient fuel for many other uses, such as cooking, refrigeration, sterilization, water heating, etc. For this reason, gas is promoted (and subsidized) by some governments and development organizations (for instance, the European Commission is funding a important gas promotion programme in the Sahelian countries of Africa). Unfortunately, at present gas, lamps can only be used in places which are easily accessible and not far from gas refilling centres; otherwise transportation costs become prohibitive.

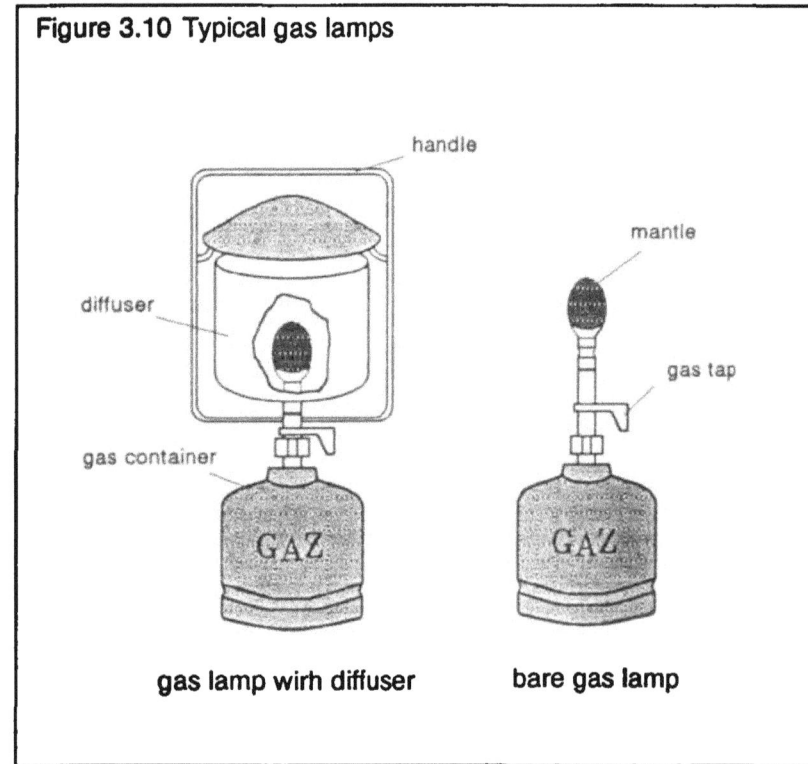

Figure 3.10 Typical gas lamps

handle

mantle

diffuser

gas tap

gas container

gas lamp wirh diffuser   bare gas lamp

Biogas lamps

Operating principles and construction

Biogas lights use rare-earth incandescent mantles in a similar way to propane/butane lamps. Biogas burns with a blue non-luminous flame and therefore requires the use of an incandescent mantle to produce light reasonably efficiently.

In a Chinese lamp, (see Figure 3.11), the gas supply is controlled by the size of the main jet and by a control valve; the gas mixes with the primary air in a venturi. The flame burns on a ceramic nozzle, and must be adjusted so that the mantle is positioned in the hottest part. The lamp will then give the best light. A reflector is fitted above the mantle in order to deflect the heat and burnt gases away from the air inlet and controls. Indian-made lamps are usually supplied with a glass shield (or globe), which protects the mantle from both draughts and insects.

Biogas is produced by a process called anaerobic digestion (see Chapter 6 for further details). Biogas contains about 60% methane (or natural gas) (CH_4) and 40% carbon dioxide (CO_2).

Biogas lights are mainly manufactured in India and China, the two countries where biogas is most widely used in rural areas. The Indian lamps are fairly expensive, because they are usually made of metal, whereas the Chinese lamps are often ceramic and are therefore cheaper, but less reliable. See Figure 3.11 which shows a metal Indian lamp and a ceramic Chinese lamp.

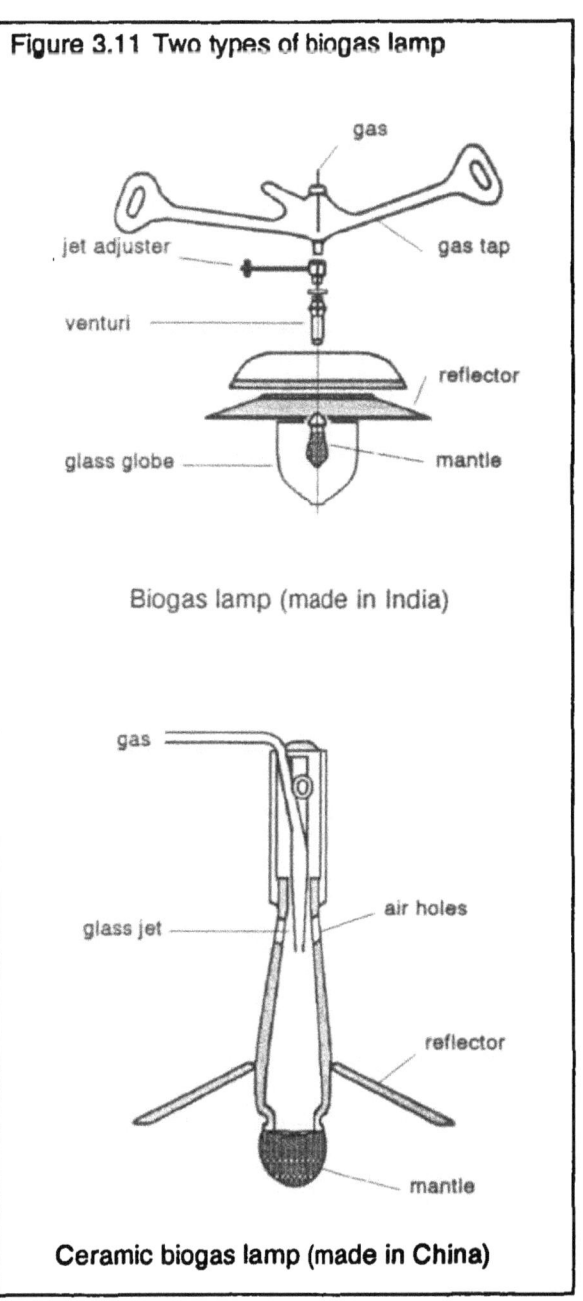

Figure 3.11 Two types of biogas lamp

Biogas lamp (made in India)

Ceramic biogas lamp (made in China)

Biogas lamps: quick data	
Colour rendering:	Poor
Luminous efficacy:	Poor (0.5 to 1 lm/W)
Luminous flux:	330 to 1300 lm
Power range:	0.10 to 2.20m³/h (biogas)
Life:	2 to 5 years, but requires spare parts (e.g. mantle)
Energy source:	Biogas
Applications:	General and localized lighting for homes.
Comment:	Quiet, fixed as opposed to portable, convenient to use, possible explosion risk, possible health hazard due to mantle.

Performance and applications

Biogas lights have a similar light output to gas lamps but typically consume between 90 and 180 (up to 900 for the very large ones) litres of biogas per hour. The large consumption of gas can be explained by the much lower calorific value of the biogas compared to methane or propane and also by the presence of impurities. Biogas lamps are reported to be expensive and to require regular servicing, but they give a good light output and are easy to use, being quiet in operation and easy to light. They can therefore provide an attractive lighting alternative in areas favourable to the use of biogas digesters.

3.4 Comparative Performance of Flame-Based Lamps

This section summarizes the main characteristics of the flame-based lighting sources. Figure 3.13 indicates graphically the relative efficacy of the different types of lamps.

Figure 3.12 Lighting of a biogas lamp

The luminous efficacy of a flame-based lighting source is the luminous flux divided by the power consumed. The power consumed is the rate of fuel use multiplied by the heating

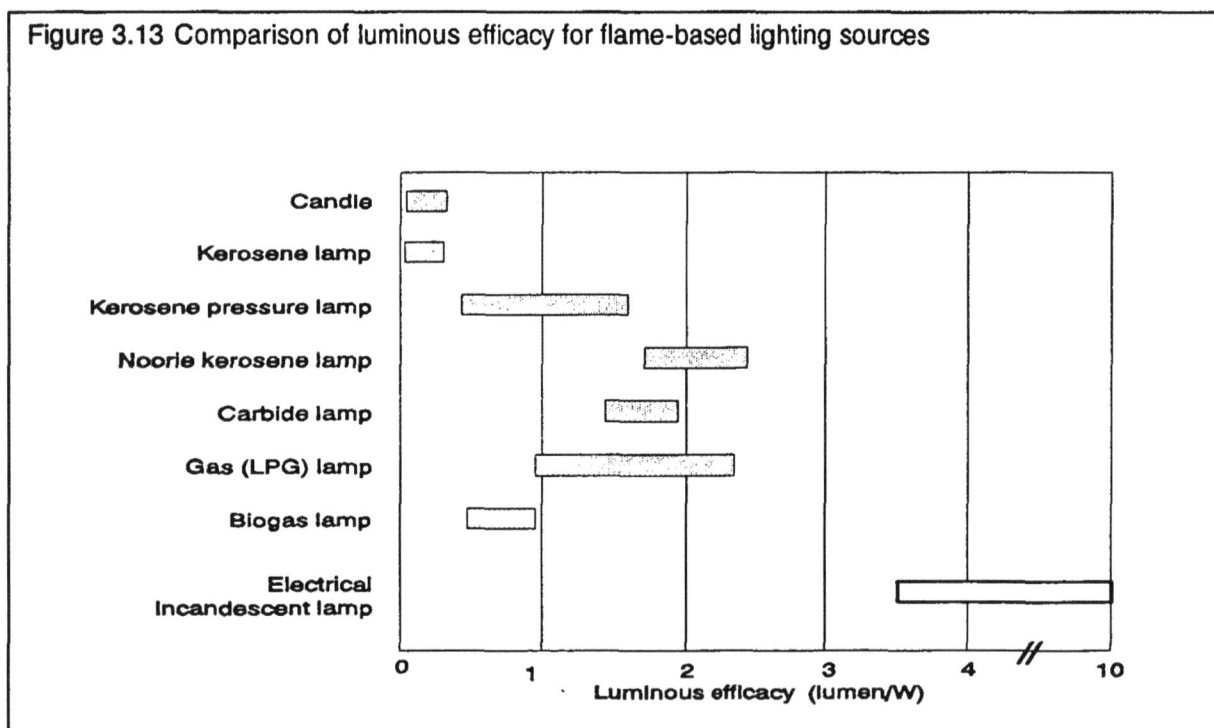

Figure 3.13 Comparison of luminous efficacy for flame-based lighting sources

Note: For comparison, the 'global' luminous efficacy (i.e. assuming a heat-to-electricity conversion of 35%) of a standard incandescent lamp which is the least efficient of all electrical lighting sources has been included on the figure.

value of the fuel. For comparison, the 'global' luminous efficacy of incandescent lamps has been included in Figure 3.13. The luminous efficacy of incandescent bulbs ranges from 10 lm/W for the standard ones up to 35 lm/W for the low-voltage halogen bulbs. However, the numbers on the figure are 3.5 to 10 lm/W respectively! This needs some explanation.

Unless electricity is produced by hydro-power plant or solar energy, electricity is produced by burning gas, gasoline, diesel or biogas either in large power plants or in small generator sets. The energy conversion from heat to electricity can be as low as 10% but can also be as high as 45%. The luminous efficacy of electrical lamps is given by electrical power consumed (see Chapter 4). Therefore, it is correct to multiply this figure by the efficiency of the heat energy to electricity conversion to be able to give a fair basis for comparing electrical light sources with flame-based lighting sources. In this case, an average heat-to-electricity efficiency of 35% has been assumed. Hence, the 'global' luminous efficacy is 10 lm/W x 0.35 = 3.5 lm/W).

Table 3.1 Performance comparison of various flame-based lamps

Type of light source	Energy source	Rate of consumption	Total power W	Luminous flux lm	Luminous efficacy lm/W	Colour rendering	Equivalent number of lamps*
CANDLE	Wax	5.50 g/h	55	1	0.02		
		7.20 g/h	72	16	0.22	Good	75
KEROSENE LANTERN	Kerosene	0.02 l/h	200	10	0.05		
		0.05 l/h	488	100	0.21	Good	12
PRESSURE LAMP	Kerosene	0.06 l/h	563	220	0.39		
		0.08 l/h	813	1300	1.60	Poor	1
NOORIE	Kerosene	0.05 l/h	513	1250	2.44	Poor	1
CARBIDE LAMP	Carbide	6.00 g/h	34	50	1.46		
		23.00 g/h	132	250	1.90	Good	5
GAS LAMP	LPG	28.00 g/h	350	330	0.94		
		34.00 g/h	425	1000	2.35	Poor	1
BIOGAS LAMP	Biogas	0.10 m³/h	693	330	0.48		
		0.20 m³/h	1385	1300	0.94	Poor	1

Examples of electric lamps for comparison

INCANDESCENT LAMP		100 W	100	1200	12	Good	1
HALOGEN LAMP		25 W	25	500	20	Good	2
FLUORESCENT TUBE		13 W	13	585	45	Good	2

*The equivalent number of lamps is the number of lamps required to produce the same luminous flux as the reference 100W incandescent electrical bulb (i.e. 1200 lm).

Either way, the energy-to-light conversion efficiency with flame-based lighting sources is much lower than with electrical lighting sources. Table 3.1 'Performance comparison of various flame-based lamps', compares the light output, fuel consumption, luminous efficacy and colour rendering of the main flame-based light sources. For example, it can be seen that 12 kerosene lanterns are necessary to obtain the same light output as a 100W standard electric bulb. However it should be noted, that considerations of capital and operating costs are of greater importance and consequence to developing country users than are the efficacy and light output of a lamp. Many people make use of hurricane lanterns or even tin-can oil lamps, because they simply cannot afford the high initial purchase price of more efficient incandescent kerosene pressure or gas lighting systems.

Electrical Lighting 4

4.1 Background and History

The idea of using electricity for lighting goes back to the early nineteenth century but it was not until the invention of an efficient vacuum pump by Sprengel in 1875 that rapid progress could be made in evacuating the glass bulbs of filament lamps. The first successful carbon filament lamps were introduced in 1878-9 by Swan in England and Edison in America. They had a luminous efficacy of only three lumens per watt and used the principle of causing a thin electrical conductor or filament to be heated to white heat in a vacuum. The vacuum both prevents oxidation of the filament and assists in maintaining its temperature. Later, tungsten replaced carbon as a more durable filament material for incandescent lamps, and it is used to this day for standard light bulbs.

Electric arc lamps had been used for lighting since 1846. These use the principle of a continuous discharge of electricity across a small air gap, much like a continuous tiny lightning discharge in a thunder storm. The so-called 'Jablochkoff Candles' were the most successful arc lamps, utilizing a narrow strip of chalk between two copper electrodes which gradually burnt away.

Research resulted in steady improvement of electric arc lamps, and in 1900 the first mercury vapour lamp was patented by Cooper-Hewitt. Discharge lamps operate on a similar principle to arc lamps except that the arc is developed in a more favourable atmosphere than air within a glass envelope or bulb.

The present day

Further improvements have led to the present generation of commercially available gaseous discharge lamps. Today the main electrical light sources used in artificial lighting are of two principal types: incandescent filament and gaseous discharge:

- Incandescent filament lamps;
    - standard incandescent tungsten lamps
    - tungsten-halogen lamps

- Gaseous discharge lamps;
    Low pressure
    - fluorescent tubes with mercury vapour
    - fluorescent compact lamps
    - sodium lamps
    High pressure
    - mercury vapour lamps
    - mercury tungsten (blended) lamps
    - metal halide lamps
    - sodium lamps.

Each of these sources will be analyzed in the following sections. The last section summarizes all the main characteristics of the different lamps.

4.2 Incandescent Filament Lamps

Standard incandescent lamps

Presently, these are the most common types of bulb in general use worldwide, and are familiar to most people. A range of common incandescent bulbs is shown in Table 4.1 at the end of the chapter. They consist of a simple filament bulb, often referred to as a GLS (General Lighting Service) bulb, and can be bought in any stores, even in developing countries. They are cheap, have good colour rendering but have a short life and are not energy-efficient in comparison with other electric lamps that will be examined.

Incandescent tungsten lamps: quick data	
Colour rendering:	Good
Luminous efficacy:	Poor (8 to 18 lm/W)
Power range:	0.75 to 1000W
Life:	15 to 1000h
Power requirements:	AC or DC, from 1.5 Volts upwards
Applications:	General, localized, local lighting for homes, classrooms, offices, etc. Should be avoided whenever possible or replaced by fluorescent lights unless the need for light is for very short periods (e.g. toilets, cupboards).
Comment:	Expensive to run because of poor energy-to-light conversion efficiency.

Operating principles and construction

In these lamps (see Figure 4.1), light is produced by the glow of a wire (the filament) heated to incandescence as a result of its resistance to the flow of electrical current. The wire is usually made of tungsten, a metal with an unusually high melting point (3350°C), low evaporation, high strength and ductility, and favourable radiation characteristics.

Most bulbs contain an inert gas, such as argon, which is there to reduce the rate of evaporation from the filament by applying a pressure to it, thereby extending the life of the bulb and preventing the glass from getting blackened quickly by metal vapour condensing on its inside surface.

Filaments can be straight, coiled, coiled coil and ribbon or flat. Coiled coil filaments are the most commonly used today, because coiling the filament reduces heat loss by means of convection in the gas filling, and results in a larger area producing more light. Incandescent lamps usually have a bayonet or screw fitting of a size which depends on their power rating.

Performance and life

The design of a filament is always the result of a trade-off between light output and life. The luminous efficacy of the filament increases with its operating temperature, but so does the rate of evaporation, resulting in a shorter life. The filament operating temperature varies from 2250°C for a 25W lamp to 2700°C for a 500W lamp. The lifetime ranges between 750 and 1000 hours on average.

Standard tungsten filament lamps convert between 8 and 12% of their input energy into light, and the remainder into heat, which is radiated into the environment. The light output decreases over the lifetime of the bulb because of tungsten evaporation, which results in reduced filament temperature and bulb-blackening. Initial luminous efficacy ranges between 8 and 18 lm/W. At the end of their life, most tungsten filament lamps provide less than 80% of their initial light output.

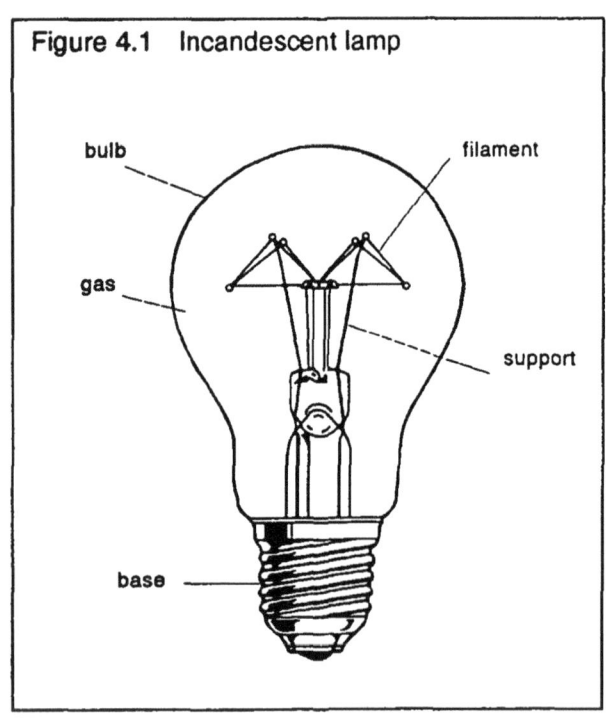

Figure 4.1   Incandescent lamp

bulb
filament
gas
support
base

34

Although the typical expected life of a standard tungsten filament lamp is 1000 operating hours, it is reduced by frequent switching on and off or other adverse conditions such as voltage fluctuations, thermal shocks (for instance from rain on the glass bulb), physical shocks and mechanical vibrations.

Lamps with a silvered reflective coating laid on the inside rear surface of the glass envelope are now available. These can be operated at a slightly reduced temperature without decreasing the light, and hence have a longer life (2000 hours). Multiple layer coatings, also called dichroic reflectors, can be used either to produce coloured light, or to reflect the light, but allow the heat to pass out through the rear of the bulb. By this means about 80% of the heat is eliminated from the reflected beam.

Tungsten halogen lamps

Halogen lamps are usually more compact than standard incandescent lamps. One of their major applications is for car headlamps because they are able to give a strong light that can be focused. They are becoming more frequently used for displays and for medical

Figure 4.2   Several types of standard incandescent bulb, including halogen lamps

applications thanks to their good colour rendering and relatively high efficiency.

Operating principles and construction

Introducing a halogen gas, such as iodine, into an incandescent lamp, reduces the evaporation rate of the filament, and consequently the blackening of the bulb. As a result, a higher filament temperature is possible, giving a 'whiter' light and increasing and stabilizing the luminous efficacy over the lifetime of the bulb. This is achieved by the halogen regenerative cycle in which evaporated tungsten atoms combine near the fila-

Tungsten halogen lamps: quick data	
Colour rendering:	Good
Luminous efficacy:	Medium (12 to 35 lm/W)
Power range:	2 to 2000W
Life:	15 to 4000h
Power requirements:	AC or DC, from 1.5 Volts upwards
Applications:	Local and task lighting for homes, offices, health centres, hospitals, workshops, torches and searchlights. Should be used to replace incandescent lamps when fluorescent lights are not appropriate or are unaffordable.
Comment:	Good for beam focusing.

ment with iodine atoms to form tungsten iodide at a temperature around 1700°C. Tungsten iodide carried by convection to the bulb wall will not adhere to it if the bulb temperature is above 200°C. Moreover, the vapour near the filament is reduced at a temperature above 2500°C to iodine vapour and tungsten, which redeposits on the filament. In order to maintain such high operating temperatures, smaller bulbs made of a strong and specially heat-resistant glass (quartz) are used. This also allows operation of the lamp at a very high pressure and hence at a higher temperature without increasing the rate of filament evaporation. Consequently tungsten halogen lamps produce more and whiter light and have a longer life than standard tungsten lamps.

It should be noted that the quartz envelopes of tungsten halogen lamps can be damaged if touched with bare fingers as the quartz absorbs finger grease which then burns when the bulb heats up. Also, since such bulbs get extremely hot, they are subject to damage from thermal shock caused by exposure to

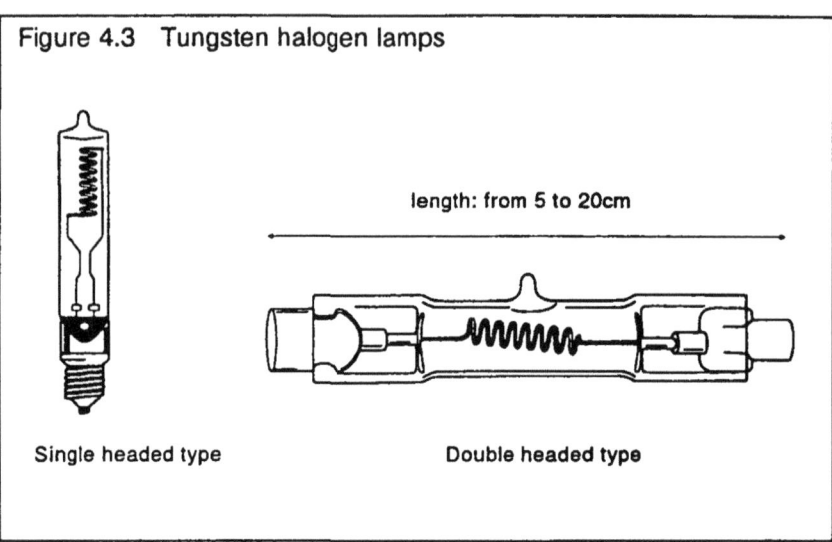

Figure 4.3 Tungsten halogen lamps

length: from 5 to 20cm

Single headed type

Double headed type

water (e.g. rain) and they can inflict serious burns if touched while operating (or while still hot after being switched off).

Tungsten halogen lamps can have different shapes, but are most often tubular, single-ended or double-ended (Figure 4.3). Their smaller size makes them suitable for good optical control as their light can readily be focused by a reflector. They are consequently widely used for motor vehicle headlamps, film and photographic slide projectors and miniature display lamps.

Low voltage (12/24V DC) closed dichroic halogen lamps are particularly appropriate for medical needs. These have an integral reflector that reduces infrared (i.e. heat) in the forward beam by up to 95%. They have a protective front glass which isolates the quartz capsule and prevents soiling of the reflector. The most outstanding features of these lamps are a very good colour rendering and the very small size of the reflector (e.g. 60mm diameter) that can be placed very close to the surgeon and to the working area without any problem (see Figure 4.4). These lamps are available with various output beam angles ranging from 8° to 60°. Figure 4.5 gives an example of illuminances at various distances from the different bulbs.

Performance and life

The lifetime of tungsten halogen lamps often exceeds 2000 hours and their luminous efficacies typically range from 12 to 35 lm/W. At the end of their life, most tungsten halogen

Figure 4.4 Low-voltage halogen lamps with faceted metal reflector and glare shield

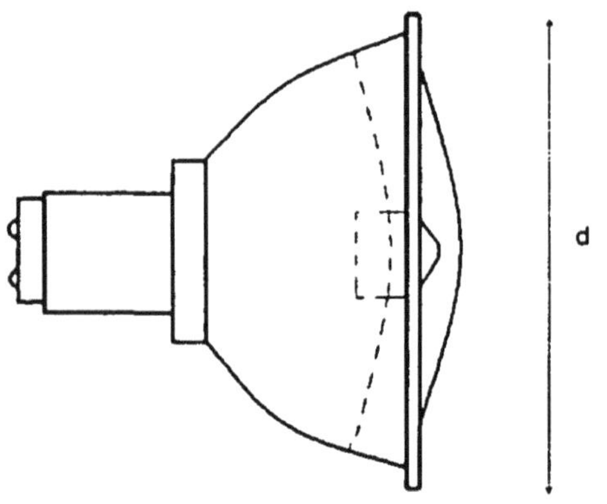

d (diameter) = 7cm for a 20 to 70W lamp

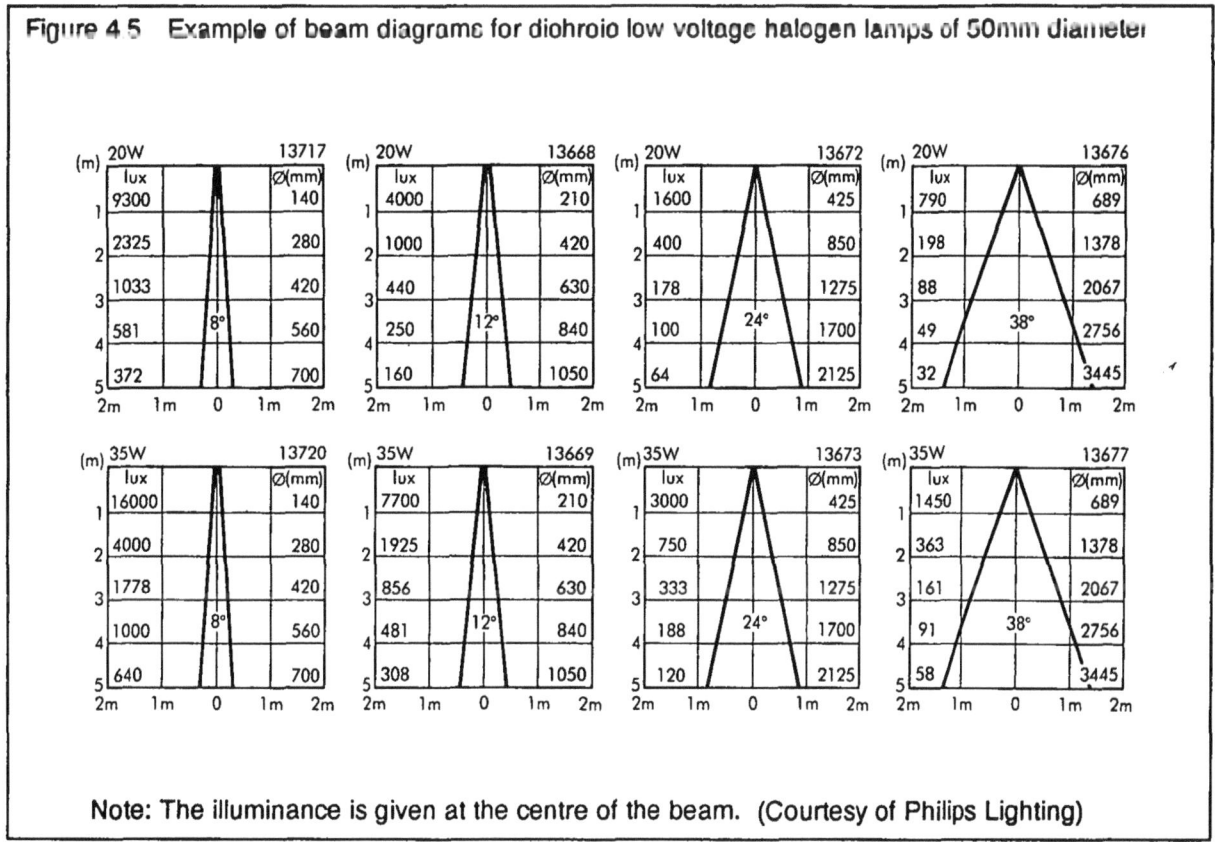

Figure 4.5 Example of beam diagrams for dichroic low voltage halogen lamps of 50mm diameter

Note: The illuminance is given at the centre of the beam. (Courtesy of Philips Lighting)

lamps will still have a light output equivalent to 95% of its initial value. The colour rendering is good to excellent.

As can be seen from above, the main advantages of tungsten halogen lamps over the standard incandescent ones are: increased light output, longer life, improved maintenance of light output over the bulb lifetime and better suitability for optical control (i.e. focused beams).

4.3 Low-pressure Gaseous Discharge Lamps

Fluorescent tubes

Fluorescent lamps, known as 'tubes' because of their shape, and standard incandescent lamps, are the two types of lamp most widely used today for a range of lighting needs. They consume on average four times less energy than incandescent lamps for the same light output, are relatively cheap, widely available and have a long life. They are however relatively bulky (e.g. 60cm for a 18W tube).

Operating principles and construction

Low-pressure mercury discharge lamps in which the inside of the discharge tubes is coated with fluorescent powders are called fluorescent tubes. The gas-discharge process begins when electrons are emitted by the cathodes placed at the ends of the tube.

Two types of cathodes exist: the hot cathodes used in pre-heated lamps, and the cold ones used in instant-start lamps. A hot cathode consists of a hollow cylinder of tungsten wire coated with electron emitting metal oxides; when a current flows through it, it is heated to incandescence (like the filament of a filament lamp) which causes free electrons to be emitted and accelerated by the applied electric field.

In lamps fitted with cold cathodes, sufficiently high voltage is applied across the lamp in order to remove electrons directly from the cathodes, heating them instantly and making them emit. In most modern lamps, called rapid-start lamps, a combination of these two mechanisms is applied. The swarm of free electrons leaving the cathode excites and

37

Fluorescent tubes: quick data	
Colour rendering:	Poor to good (depends on the chemicals inside)
Luminous efficacy:	Good (35 to 78 lm/W)
Power range:	4 to 125W
Life:	5000 to 8000h
Power requirements:	AC with control gear, from 100V upwards
	DC low voltage (from 3V) with inverter/transformer.
Applications:	General, localized and task lighting for homes, offices, health centres, hospitals, workshops etc. Should replace incandescent lamps wherever light is needed for lengthy periods of time. Never for short periods of lighting.
Comment:	Provide a cost-effective way for lighting when energy is scarce or expensive. Should be disposed of carefully to avoid mercury pollution.

ionizes argon atoms present in the tube, which in their turn excite and ionize mercury atoms. The excited mercury atoms are not stable; they revert immediately to their initial state, and doing so they release photons (ultra-violet), which are converted into visible radiation by the fluorescent powders that coat the inside of the tube (see Figure 4.6)

Fluorescence is a physical property of certain substances by which they radiate visible light when irradiated with invisible radiation such as ultraviolet light. These powders, called phosphors, are made of semiconductor materials to which activators have been added.

Early phosphors were made of zinc beryllium silicate with silver as an activator, but because of the toxicity of beryllium, present lamps use mixtures of calcium halophosphate and magnesium tungstate.

Mercury vapour is included in the atmosphere inside the tube because it has a very high ratio of converting UV radiation into visible radiation and thereby it adds to the efficacy of the lamp. Argon is necessary to help establish the arc, to ionize vaporised mercury and to contain the mercury arc.

Ballasts and circuits for AC systems

As an arc discharge has a negative resistance characteristic (i.e. the resistance tends to decrease as the current increases) it is necessary to place a current-limiting device in series with it or the current would simply build up and either damage the lamp or cause the circuit to trip or blow a fuse. Lamps use reactive (magnetic) and, more recently, electronic ballasts.

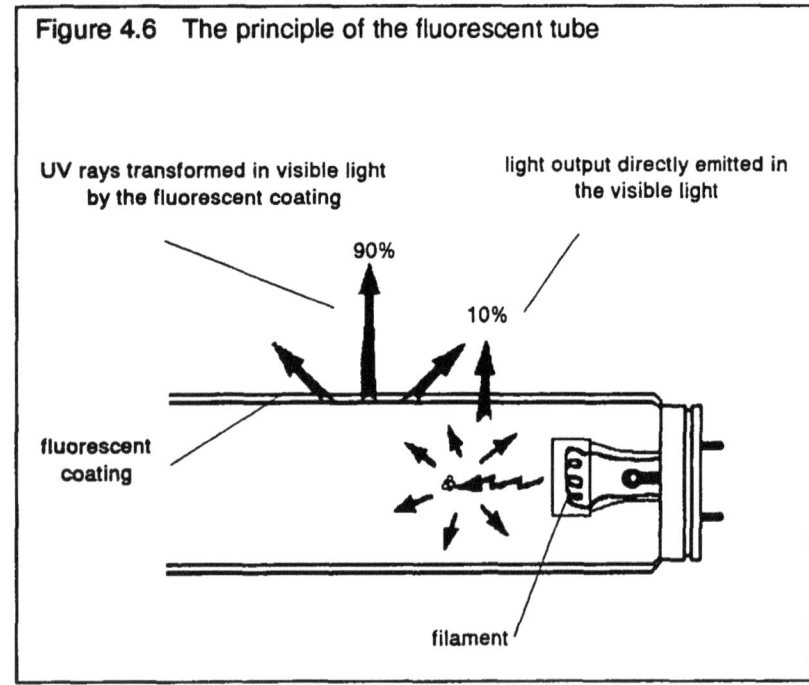

Figure 4.6 The principle of the fluorescent tube

UV rays transformed in visible light by the fluorescent coating

light output directly emitted in the visible light

90%

10%

fluorescent coating

filament

Figure 4.7 Small fluorescent tubes. From the bottom; 4W, 6W, 8W fluorescent tubes. The 8W tube is fitted in a 12V luminaire

Figure 4.7 Small fluorescent tubes. From the bottom; 4W, 6W, 8W fluorescent tubes. The 8W tube is fitted in a 12V luminaire

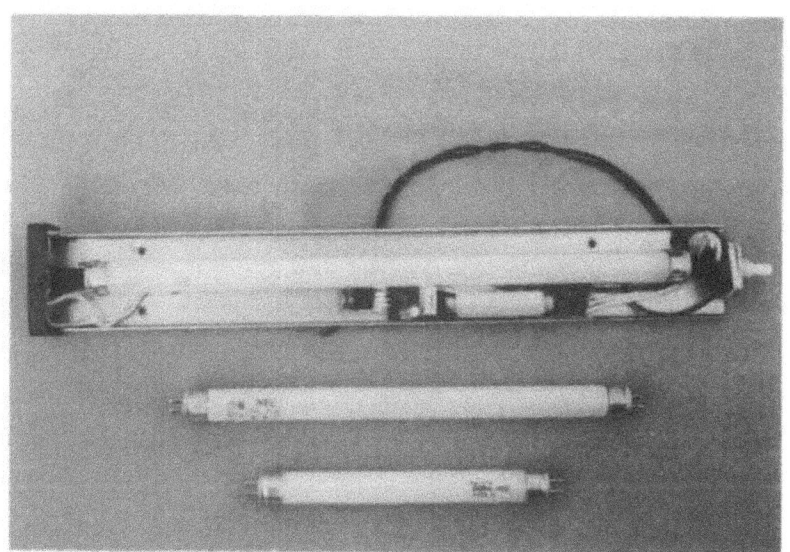

Ballasts have the following functions:

- Limiting the current in the negative resistance arc;

- Providing sufficient voltage to initiate the arc (at least 100V);

- Regulating the current against voltage variations;

- Heating the cathodes in pre-heat and rapid-start lamps;

- Providing a high power factor.

Until recently most ballasts were based on an induction coil which uses the principle of a transformer, or the ignition coil in a car, to produce the high voltage needed to strike the arc. More recently, electronic ballasts are becoming popular as they can supply high-frequency current (typically 30kHz), which eliminates all stroboscopic and flickering effects. This also increases the luminous efficacy and life and tends to lead to a quicker establishment of the arc when switching on.

There are two main types of starting circuits: switched and switchless. Figure 4.8 shows the operation of a switch-type starter. The starter consists of two bimetallic electrodes enclosed in an evacuated glass bulb.

When the master switch is turned on, an arc takes place between the electrodes and heats them, making them bend toward each other until they touch. Current is then allowed to flow through and heat the main tube electrodes, ionizing argon around them. The starter electrodes cool down and spring apart after a few seconds. Because of the choke in the circuit, there is an inductive voltage surge across the lamp, which makes the main arc strike.

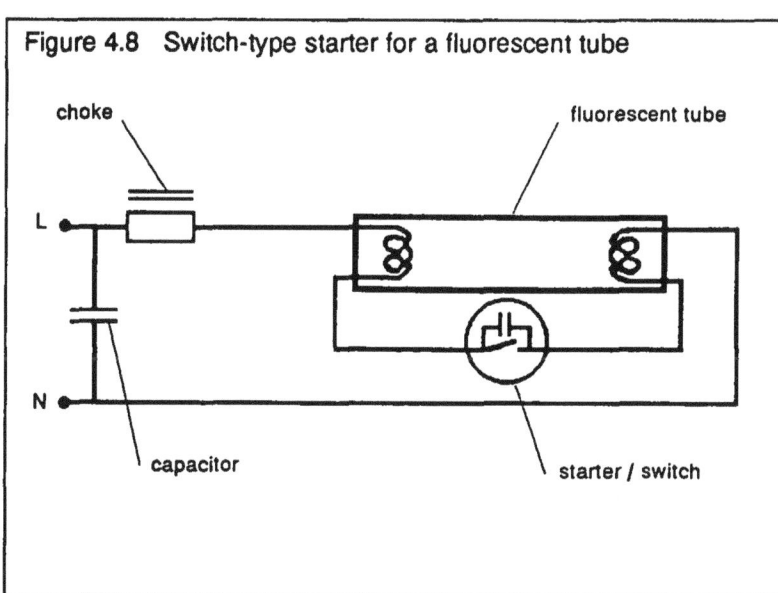

Figure 4.8 Switch-type starter for a fluorescent tube

choke

fluorescent tube

L

N

capacitor

starter / switch

Quick-start and semi-resonant start are switchless circuits employing a cathode-heating transformer wired in series with the choke. Earthing is necessary to provide a capacitive effect, which helps the arc strike.

Electronic starters use semi-conductors to produce the same effect as switch-type starters, but without any moving parts. With some high frequency electronic ballasts, it is possible to reduce or increase the light output to allow dimming.

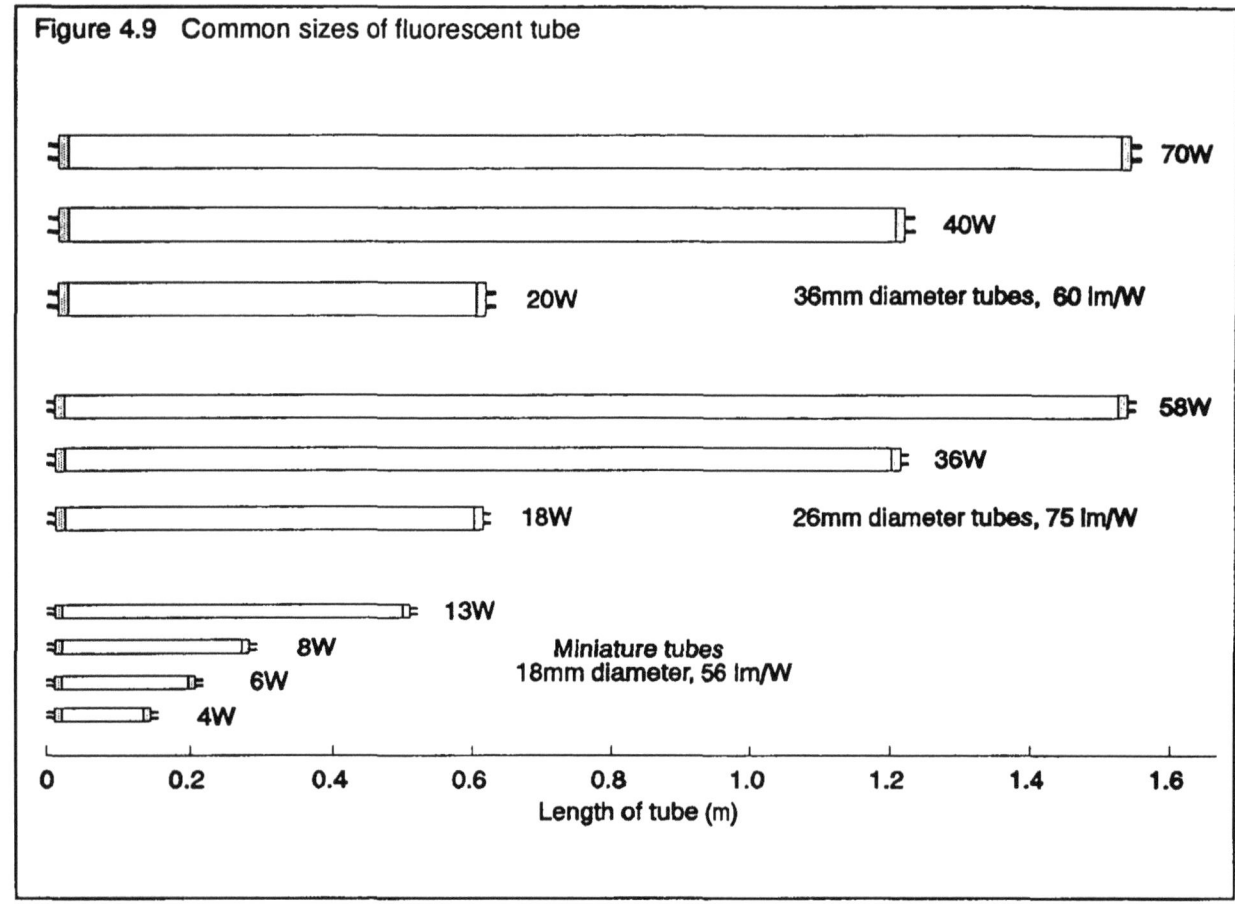

Figure 4.9 Common sizes of fluorescent tube

70W

40W

20W   36mm diameter tubes, 60 lm/W

58W

36W

18W   26mm diameter tubes, 75 lm/W

13W

8W   *Miniature tubes*
           18mm diameter, 56 lm/W

6W

4W

0   0.2   0.4   0.6   0.8   1.0   1.2   1.4   1.6

Length of tube (m)

Fluorescent tubes in DC low-voltage systems

Because fluorescent tubes need a voltage higher than 100V, an inverter is necessary to transform low DC voltage (e.g. 12V or 24V) to the required AC voltage. It is a common mistake to believe that tubes which work with a DC supply are special: they are exactly the same as those manufactured for mains electricity.

The inverters are designed specifically to start as well as to operate the tubes, hence they should not be used for other applications. They are usually fixed inside the luminaire housing.

Inverter

The inverter is the most expensive component of a complete luminaire (i.e. tube, inverter, fittings, housing). If the price of a complete luminaire is very low, it usually means that the inverter is of bad quality. It is then likely that the light output might be lower than expected, the tube will have a shorter lifetime and the power consumed by the inverter will be very high. Indeed in some cases, the power consumed by the inverter may be higher than that consumed by the tube. This can also happen when the inverter and tube are mismatched, for example, an inverter designed for a 20W tube powering an 8W tube.

Therefore, when choosing a DC fluorescent luminaire, it does not matter if the diffuser is made of expensive and attractive material (in any case, it may need to be removed after a few days of use because insects may nest inside) but it is important to choose a good quality inverter.

Good quality inverters as opposed to low quality ones are often reparable because they are made of common electronic components that are available or can be ordered from most radio repair shops, even in developing countries. Finally, it is worth having integrated protection against the possibility of polarity reversal to avoid damaging the electronics.

Performance and life

Of the input energy to a fluorescent tube, only about 22% ends up as light; 36% is converted into infrared radiation and 42% into heat. The luminous efficacies of fluorescent lamps range from 35 to 100 lm/W which is four times better than for standard incandescent lamps. The lifetime of fluorescent lamps varies from 5000 to 8000 hours, at least four times better than standard incandescent lamps.

The light output of fluorescent lamps can be greatly increased by reducing the tube diameter. Using improved gas mixtures and phosphor coatings can also increase output. Recently, manufacturers have been moving towards 26mm diameter tubes (instead of 38mm standard tubes) with three (triphosphor) or more layers of rare earth phosphor coating and a krypton gas filling.

Frequent switching on and off, voltage fluctuations (undervoltage and particularly overvoltages) and variations in ambient temperature larger than 5°C around the standard value of 20°C will result in lower light output, a shorter lifetime, and poorer colour rendering.

There exists a wide power range in which fluorescent tubes are available. The smallest fluorescent tubes are about 4W and 0.15m long. These are used in small portable lamps. The most powerful are about 125W and 2.40m long and are for general lighting.

Compact fluorescent lamps

The introduction of compact fluorescent lamps (CFLs) in the early 1980s signalled the start of an energy-efficiency revolution in lighting, as CFLs can often simply be substituted for conventional incandescent filament bulbs and in any case they do not need the large and often unsightly fittings needed for conventional fluorescent tubes. The smaller CFLs are about the size of 100W standard incandescent bulbs.

Operating principles and construction

Compact Fluorescent Lamps (CFLs) work in the same way as standard fluorescent tubes, the main difference being that CFL tubes are smaller and the tubes are folded over and sometimes mounted in pairs.

Two main CFL types exist:

- *Integral* : with the magnetic or electronic ballast built into the base of the lamp and a standard BC (bayonet) or ES (Edison screw) adapter to fit conventional bulb sockets;

- *Modular* : where the ballast is incorporated in a separate fitting or 'adapter' with a socket specially designed to take a detachable folded tube assembly. In most cases, the adapter has a BC or ES base, so that the combination of adapter and tube can be plugged into any standard lighting socket.

Compact fluorescent lamps: quick data	
Colour rendering:	Good
Luminous efficacy:	Good (48 to 80 lm/W)
Power range:	9 to 23W for integral CFLs, 5 to 55W for modular
Life:	8000 to 10 000h
Power requirements:	AC (no need of control gear for the integral type), from 100V upwards. DC low voltage (from 12V upwards) with inverter/transformer.
Applications:	General, localized and task lighting for homes, offices, health centres, hospitals, workshops, etc. Never for short duration lighting (e.g. toilet, cupboards, etc.).
Comment:	Offer major energy savings when replacing incandescent standard bulbs. A very good choice indeed for most lighting needs.

Figure 4.10 Various types of compact fluorescent lamp

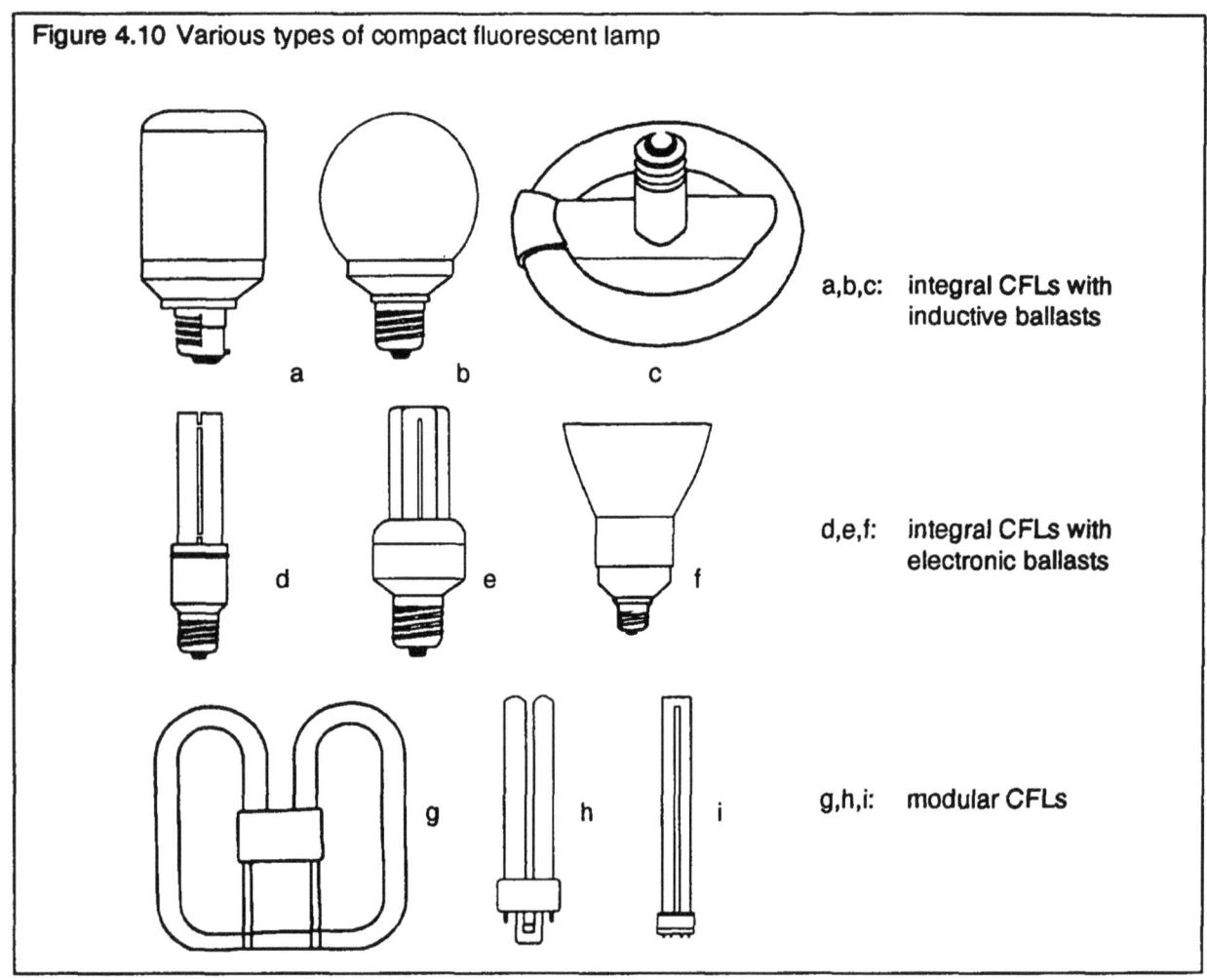

a,b,c: integral CFLs with inductive ballasts

d,e,f: integral CFLs with electronic ballasts

g,h,i: modular CFLs

It can be seen from Figure 4.10 that apart from the most commonly implemented form, of twin folded tubes sticking straight out from the adapter or holder like 'fingers' from a hand, there are a few alternative forms, including the so called 'D' tube (where a single slim tube is wrapped around its adapter in a squarish shape. There is also a type with a circular fluorescent tube (these are actually not all that compact and have been around for some time) and several forms with a glass outer globe or diffuser surrounding a 'finger'-type CFL unit.

The choice is complicated further because most manufacturers of both integral and modular CFLs offer a choice of inductive or electronic ballasts. The electronic ballast is superior to the inductive ballast in several ways:

● More energy-efficient (i.e. uses slightly less power);

● Strikes its arc more quickly;

● Starts virtually instantly and warms up to full light emission levels more quickly;

● Significantly lighter in weight and even more compact;

● Has a better power factor (close to 1) than the inductive type.

The last advantage, concerning the power factor, needs some explanation. Inductive ballasts have an inherently poor power factor, and cause a large phase difference between current and voltage. This, in turn, can undermine some of the economies compared with a tungsten filament lamp, by creating an extra power loss in the transmission line.

Fortunately for the user, these line losses are not measured by their electricity meter, so

Figure 4.11 Compact fluorescent lamps (20W) compared with a 100W bulb

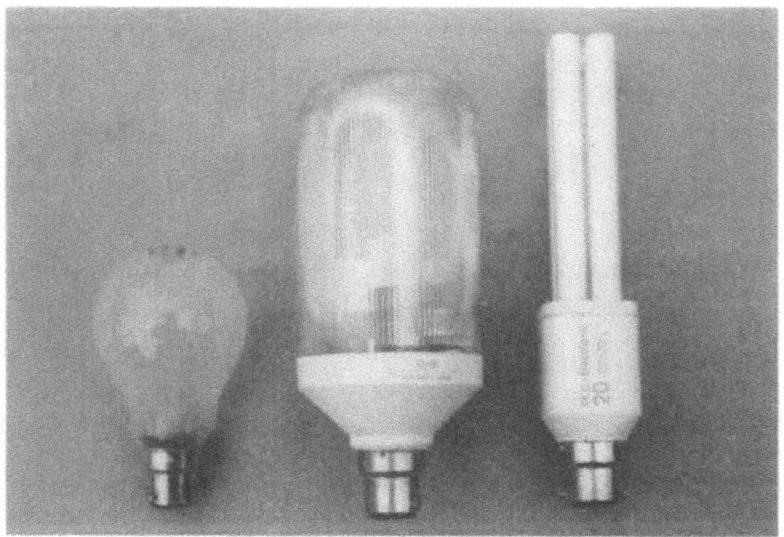

Performance and life

CFLs are attractive because they consume only one-quarter to one-third as much electricity as the incandescent lamps which they can so easily replace, while offering similar light output. They also last five to eight times longer; e.g. typically 8000 to 10,000 hours. The luminous efficacy ranges from 45 to 60 lm/W for the integral CFLs and for the modular CFLs when including the control gear.

Modular CFLs offer the advantage that the ballasts do not have to be replaced with the lamp. While CFL tubes last 8000 to 10,000 hours, ballasts may last 50,000 hours or

they get the full financial benefit as if the line losses did not exist. However, if a large proportion of the electricity users in a country opted for inductive rather than electronic CFLs, the benefit in reducing the national demand for electricity would be somewhat less than might be expected from the national improvement in efficacy.

Inductive ballast CFLs that were manufactured in the early 1980s are being gradually replaced in the market-place by electronic ballast units. The only disadvantage of electronic ballasts is that, at present, they tend to cost more than the inductive type. However, in time, as more units are sold, the electronic type looks set to come down in price and then it seems likely to become the standard form for CFLs, by virtue of its many other advantages.

Many manufacturers of CFLs now offer small reflectors (see Figure 4.10, lamp f) that can be clipped to the base of the integral CFLs and that reflect the light emitted from the sides of the tube so that a beam is formed in the direction that the bulb is pointing. This makes them much more effective for direct and local lighting.

An alternative is to use the 'D' type CFLs (Figure 4.10, lamp g) which radiate light primarily away from the wall or ceiling to which they are fitted. They are generally used concealed within a translucent luminaire which diffuses the light output.

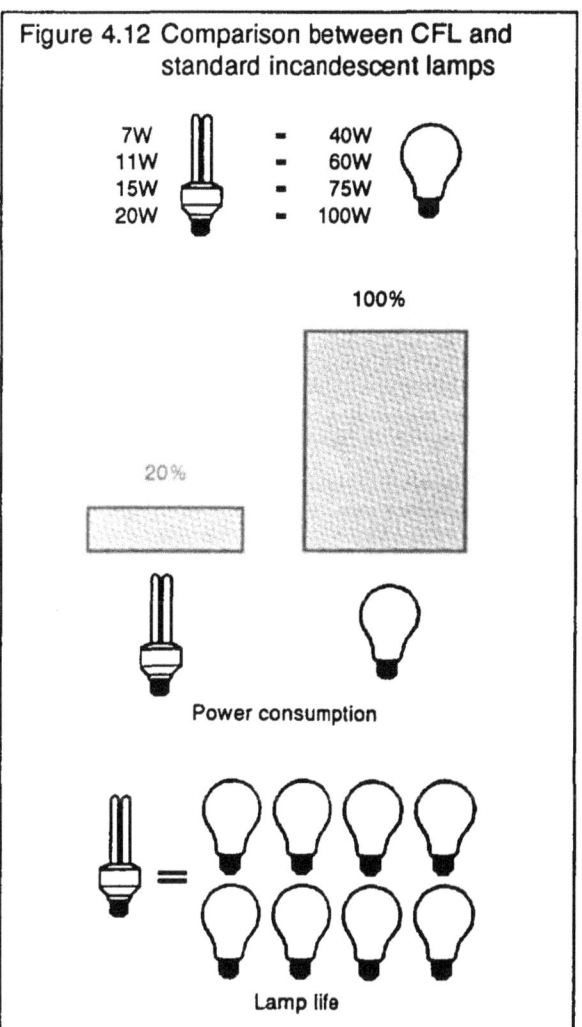

Figure 4.12 Comparison between CFL and standard incandescent lamps

7W	=	40W
11W	=	60W
15W	=	75W
20W	=	100W

100%

20%

Power consumption

Lamp life

more. Therefore it makes economic sense to use the modular system where the shorter-lived and relatively less expensive tubes can be replaced on their own.

Many modular (i.e. separate) ballasts and adapters are often purpose-built for a specific brand and power rating of CFL tube, and the socket for the tube is deliberately non-standard so that it is impossible to fit another manufacturer's product (or even a different wattage CFL tube from the same manufacturer). Some standardization is developing to allow substitution of different CFL tubes; care needs to be exercised to check what replacements can be used with a modular system, and whether these can be readily obtained.

Most CFLs will not strike effectively at sub-zero temperatures. Therefore CFLs are not suitable as exterior lights in cold climates for this reason. Potential users should check the manufacturers' specifications which normally warn of this where relevant. CFLs, like all fluorescent lamps, only radiate at their rated light output once they are fully warmed up. In cold conditions they can be noticeably dim for several minutes after switching them on. Also they are not well suited to applications where they need to be switched on and off very often (for example in a WC or a storage cupboard) but are ideal for use in places where lights need to be on more or less continuously (and where the maximum benefit can be gained from their economic use of electricity).

In spite of the relatively high price of CFLs, which seems to be partly due to the small volume of production, they are still often more cost-effective than standard incandescent lamps because of their very low power consumption. Furthermore it is to be hoped that as they get cheaper, they will sell in larger quantities and the resulting economies of scale in production will be passed on to the consumer, making CFLs even more cost-effective. CFLs manufacturers are currently manufacturing lamps that could be widely used in developing countries where the tech-

Figure 4.13 The latest generation of compact fluorescent lamps (Courtesy of Philips Lighting)

nology must adapt to more demanding operating conditions (e.g. larger fluctuations of voltage [up to 10%] and temperature).

Large-scale introduction of CFLs

Several projects to help introduce CFLs on a large scale are already underway or just starting in various countries, including the developing countries, for example Brazil, Pakistan, Egypt, India and Mexico. These projects are intended to replace anything from several thousands to up to a few millions of incandescent bulbs by CFLs. They are usually supported by electric utilities together with international aid agencies and in some cases involve CFLs manufacturers. Some of the projects are designed to overcome market resistance to the high first-cost and unfamiliarity of CFLs and are based on the principle of leasing. The project leases CFLs to its customers for a small monthly fee (e.g. US$0.25). The lease payments, spread over a few years, will repay the interest-free cost of the CFLs. Each month the CFLs save consumers more in avoided energy bills than they cost in lease payments. Over their eight-year life the CFLs will save the economies of the countries involved several times (typically over five times) their purchase price in avoided capital investment for peak power generation. Furthermore, the electricity saved by the CFLs in urban areas can be made available to semi-urban regions that currently suffer from power cuts during evening peak hours.

Low-pressure sodium lamps: quick data	
Colour rendering:	Very poor (gives a monochromatic yellow light)
Luminous efficacy:	Excellent (100 to 200 lm/W)
Power range:	8 to 180W
Life:	6000 to 24 000h
Power requirements:	AC from 100V upwards, DC low voltage (from 12V upwards) with inverter/transformer.
Applications:	Out-door lighting: street lights, roads
Comment:	Thanks to their excellent energy efficiency, they can be used for photovoltaic outdoor lighting

Low-pressure sodium discharge lamps

These lamps are easy to recognize because of their very distinctive yellow light. They are mainly used for street and road lighting.

Operating principles and construction

In low-pressure sodium lamps the light is produced from an arc which is established in a U-shaped tube containing sodium with a small amount of argon and neon to assist in striking the arc. The pressure inside the arc tube is very low; almost a vacuum at about 1/1000mm of mercury pressure. The visible light produced is almost monochromatic yellow light at wavelengths of 589nm (95%) and 586nm (5%); i.e. it is emitted over a very narrow range of wavelengths.

The arc tube is placed inside an outer tube, and the space between the two is under vacuum (Figure 4.14). The arc tube is made of special glass with a high resistance to the attack of sodium vapour. It operates at a temperature about 260°C. The enclosing tube is made of ordinary glass. Its function is to protect the arc tube and maintain it in a constant thermal environment.

The lamp casing has a tubular shape and some lamps can only be operated in a horizontal or a vertical position. They are generally available in the following power ratings: 18, 35 and 55 watt lamps operate in upright or horizontal burning positions, but 90, 135 and 180 watt lamps are intended for a horizontal burning position only.

The starting voltage is 600V, which is provided by a leakage transformer acting as a choke after start-up. A power-factor capacitor is included in the circuit. The starting time of low-pressure sodium lamps may last from 9 to 15 minutes while the lamp warms up and vaporizes the sodium (glowing a dull red colour initially); restriking will take only 30 seconds once warm.

Performance and life

In low-pressure sodium lamps 35.5% of the total energy input ends up as visible light, the

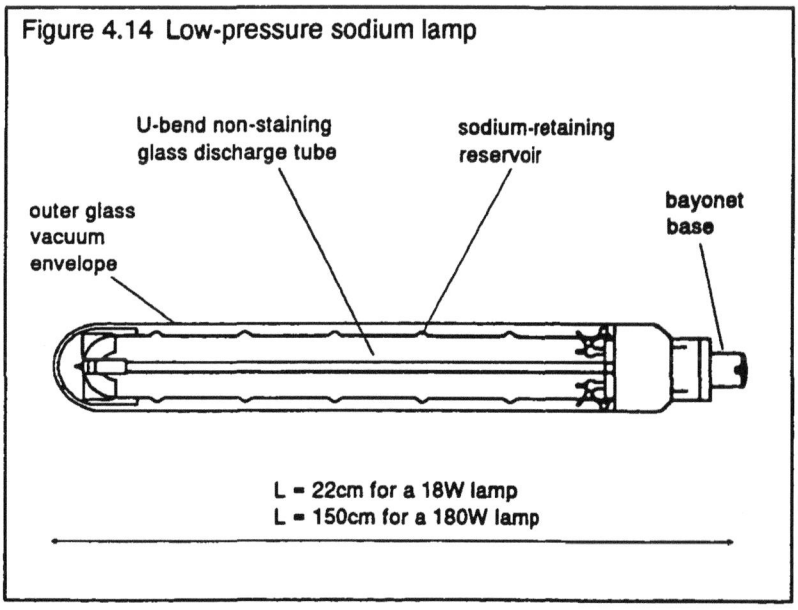

Figure 4.14 Low-pressure sodium lamp

U-bend non-staining glass discharge tube

sodium-retaining reservoir

outer glass vacuum envelope

bayonet base

L = 22cm for a 18W lamp
L = 150cm for a 180W lamp

remainder being converted into heat (60%) and infrared radiation (4.5%). These lamps have the highest efficacy of any type of lamp: from 100 to 200 lm/W, and have a lifetime from 6000 to 24,000 hours.

The light produced is an almost monochromatic yellow light, which has a very poor colour rendering. Consequently, these lamps are used mainly for road and street lighting where the monochromaticity is not a serious disadvantage. In addition, the yellow light has the property of penetrating fog and mist better than white light, while tending not to dazzle and cause glare to the same extent as white lights. These lamps are often used with photovoltaic-powered street luminaires because of their high luminous efficacy.

4.4 High-pressure Discharge Lamps

These lamps are not generally available on the shelves like most of the other types described previously. They are designed mostly either for large commercial and industrial needs or for other particular lighting requirements (e.g. photography, filming). Even for the smaller lamps of this type, the light output is very high compared to the requirements for most lighting situations in rural areas.

They are designed to run on a stable and reliable AC supply and consume quite a lot of power. Consequently they are not suitable for most applications in rural areas. However it is of importance for the reader to know about them in case of special needs in semi-urban or semi-industrialized areas.

The high-pressure discharge sources are:

● Mercury vapour lamps;

● Mercury tungsten lamps (blended);

● Metal halide lamps;

● High-pressure sodium lamps.

These categories of lamps are also referred to as high-intensity discharge (HID) lamps.

Mercury vapour lamps

These lamps are used for industrial and commercial applications and require rather complicated control gear and fittings.

Operating principles and construction

The light-producing element of a mercury vapour lamp is an arc tube containing mercury and small amounts of argon, neon and krypton. The arc tube is made of quartz, which must withstand an arc temperature of 1000°C. It is placed inside an outer tube made of borosilicate glass. The space between the two tubes is filled with nitrogen, which provides the arc tube with thermal insulation, and protects metal parts from oxidation. The arc tube operates at a pressure ranging from 1 to 10 atmospheres. Figure 4.15 shows a typical construction of a mercury lamp.

The arc tube contains two main electrodes and one starting electrode. Each main electrode is made of a double layer of coiled tungsten wire supported by a tungsten rod, and dipped into a mixture of oxidized thorium, calcium and barium carbonates.

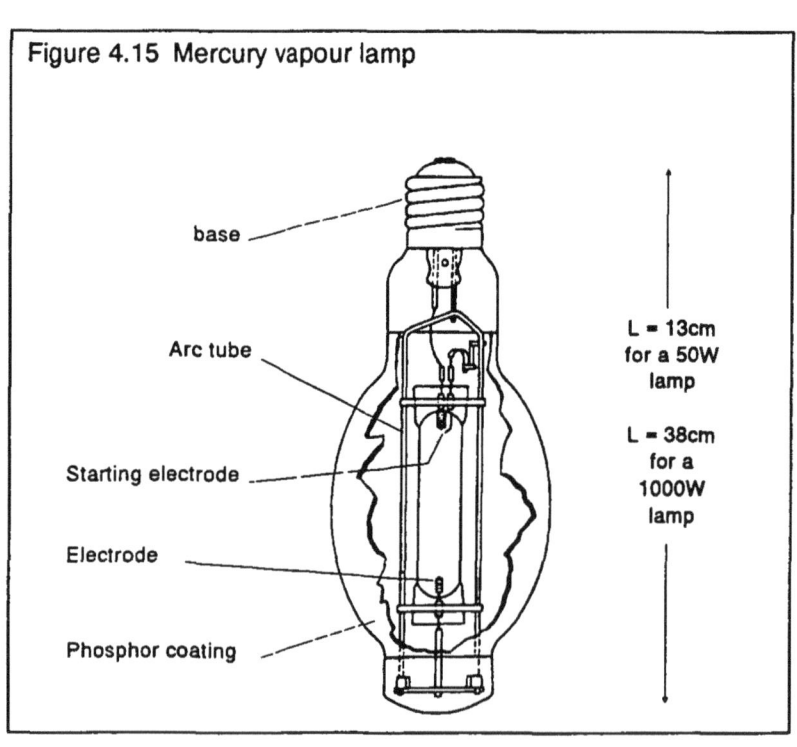

Figure 4.15 Mercury vapour lamp

base

Arc tube

Starting electrode

Electrode

Phosphor coating

L = 13cm for a 50W lamp

L = 38cm for a 1000W lamp

Mercury vapour lamps: quick data	
Colour rendering:	Poor to good (depends on chemicals inside)
Luminous efficacy:	Good (38 to 63 lm/W)
Power range:	50 to 1000W
Life:	5000 to 24 000h
Power requirements:	AC from 100V upwards, need of control gear
Applications:	Industrial and commercial lighting (e.g. factories, offices)

When voltage is applied to the lamp, an argon arc is struck between the starting and adjacent main electrodes. The heat generated by this local arc vaporizes mercury, and soon the main arc jumps between the two main electrodes, while the starting arc ceases. Blue-green visible light is produced and some UV radiation. While mercury continues to vaporize, the light output increases and reaches its maximum after five to seven minutes.

The UV radiation can be converted into visible light by a phosphor coating placed inside the outer tube, both increasing the light output and improving the colour rendering of the lamp. The outer tube has several important functions. It filters the UV radiation, preventing skin and corneal burns (in present lamps, the arc tube is disabled as soon as the outer tube is broken). It also provides the arc tube with a constant thermal environment, and provides a surface for the phosphor coating.

Lamps come in several shapes: bulged tubular, tubular, elliptical, pear, straight, parabolic and with integral reflectors.

Performance and life

On average, a mercury vapour lamp will convert only 15% of the total input power into light, the remainder ending up as infrared radiation (15%) and heat (70%). As a result, these lamps have a relatively poor efficacy as compared to other HID lamps. Luminous efficacies increase with lamp wattage from 38 up to 57 lumen per watt for clear-bulb lamps, and from 42 up to 63 lumen per watt for phosphor-coated lamps. Mercury vapour lamps typically have a lifetime of 24,000 hours. The colour rendering, which can be good to excellent, varies a great deal depending on the chemicals used (e.g. special phos-

phors) and it is therefore essential to consult a manufacturer.

Other disadvantages are poor lumen maintenance over the lifetime of the bulb, and difficult optical control (because of the outer tube phosphor coating). Light control can be achieved with a combination of prismatic lens and reflector, but this will then slightly decrease the light output due to transmission and reflection losses.

Blended lamps

These lamps are suitable for large lighting installations (e.g. halls, large offices) and can be used as a replacement for large incandescent lamps (e.g. 250W to 500W).

Operating principles and construction

Light is produced by an electric discharge in a high-pressure mercury atmosphere contained in an arc tube in series with a tungsten filament heated to incandescence. These are contained within a glass envelope with a fluorescent coating. The resulting light output is a combination of the light characteristics of an incandescent and mercury lamp. There is no need for control gear. The lamp gives some light output immediately when switched on, with a warm-up period of about four minutes to achieve 90% of full light output.

Performance and life

These lamps have a relatively poor efficacy as compared to other HID lamps, initial efficacies varying from 15 to 35 lm/W. Their lifetime, in average 4000 to 6000 hours, is limited by the failure of the filament. These lamps are often used as a replacement for

Blended lamps: quick data	
Colour rendering:	Good
Luminous efficacy:	Medium (15 to 35 lm/W)
Power range:	100 to 500W
Life:	4000 to 6000h
Power requirements:	AC from 100V upwards. No need for control gear. DC low voltage (from 12V upwards) with inverter/transformer.
Applications:	Industrial and commercial lighting (e.g factories)
Comment:	Unlike other HID lamps, provides light immediately after switching on.

tungsten filament lamps where lamp life is important, e.g. because access is difficult.

Metal halide lamps

These lamps are used for many applications ranging from industrial lighting needs to the lighting of movie scenes.

Operating principle and construction

With metal halide lamps, light is produced by an arc tube, which has the same construction and operating principles as that of the mercury vapour lamp. However, besides mercury, argon, neon and krypton, the arc tube of a metal halide lamp contains halide salts (iodides) of metals such as mercury, sodium, scandium, thallium, indium and caesium. These halides add missing red, orange and yellow colours to the typical mercury blue-green colours, hence balancing the colours of the lamp across the spectrum of colours.

Other differences from mercury lamps are the addition of white reflective coatings to the ends of the arc tube in order to maintain a more uniform temperature across it, and the inclusion of a bimetal switch, which shorts the starting electrode to the main electrode start-up in order to prevent the build-up of small electric fields

between them. The process from start-up to full light output may take up to five minutes.

No phosphor coating is placed inside the outer tube and therefore the light can be easily controlled by optical reflectors. The outer tube is made of ordinary glass, which acts as an ultraviolet filter, while maintaining the arc tube in a constant thermal environment.

Different lamp shapes exist: bulged tubular, tubular, elliptical and ellipsoidal. Some of them are meant for vertical burning, others for horizontal burning position. Compact-source metal halide lamps have been developed for projection purposes as in films and television as well as for high-power floodlighting.

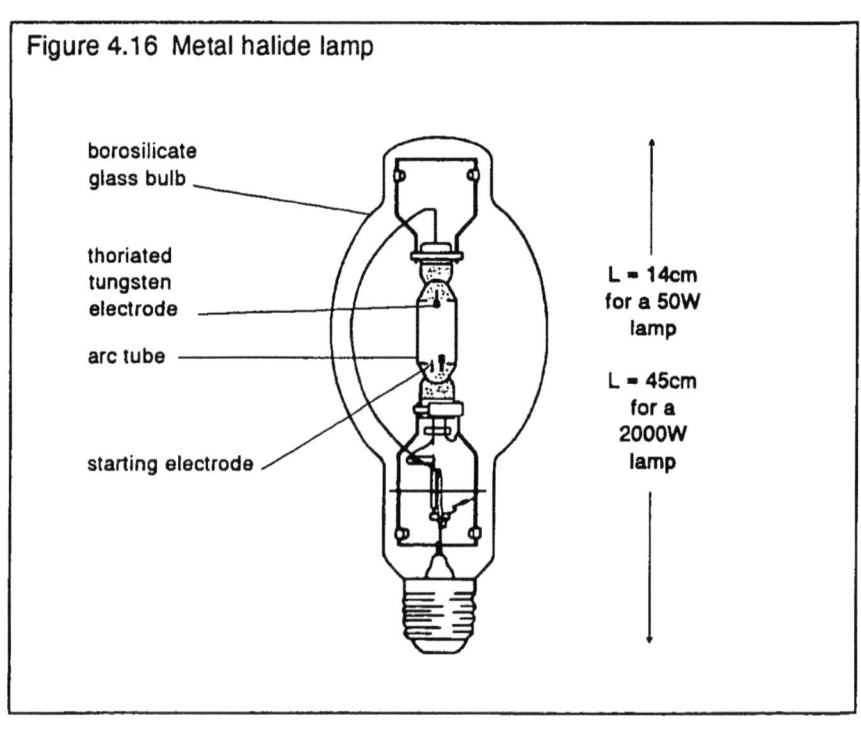

Figure 4.16 Metal halide lamp

borosilicate glass bulb

thoriated tungsten electrode

arc tube

starting electrode

L = 14cm for a 50W lamp

L = 45cm for a 2000W lamp

Metal halide lamps: quick data	
Colour rendering:	Good to excellent
Luminous efficacy:	Good (80 to 120 lm/W)
Power range:	70 to 3500W
Life:	1500 to 15 000h
Power requirements:	AC from 100V upwards, need for control gear
Applications:	Industrial and commercial lighting (e.g. displays)

Performance and life

Of the total input energy to a metal halide lamp, on average 24% ends up as light, the remainder being converted into heat (51%) and infrared radiation (25%). The addition of the metallic iodides significantly increases the light output and luminous efficacy of metal halide lamps in comparison with those of mercury vapour lamps, from 80 to 120 lm/W. However, at the same time, it requires a higher voltage both for start-up and re-ignition of the arc, increasing electrode deterioration. As a result, metal halide lamps have a relatively shorter life (1500 to 15,000 hours) and a poorer lumen maintenance over their lifetime than other HID lamps.

High-pressure sodium lamps

These lamps are similar in shape to high-pressure mercury lamps and in general have the same applications (e.g. large industrial and commercial buildings) but have a higher luminous efficacy.

Operating principles and construction

The light-producing element in a high-pressure sodium lamp is similar to that of a mercury lamp, with the following exceptions:

- The tube is made of poly-crystalline alumina (PCA), which is resistant to sodium attack;

- The arc tube is filled with xenon gas, and contains a mixture of mercury and sodium;

- The arc tube contains only two main electrodes, and no starting electrode;

- The outer tube is made of heat-resistant glass; the space between the outer and inner tube is evacuated.

Because of the absence of a starting electrode, the ballast must provide a much higher starting voltage than in the case of other HID lamps. This is achieved by means of a superimposed pulse on the ballast open-circuit waveform. This pulse produces a voltage of about 2500V for the very short time necessary to ionize the xenon gas, so that the open-circuit voltage can strike the xenon arc. The initial light is produced by the excited xenon and mercury, and has a bluish-white colour. As the temperature rises, sodium is excited and produces the low-pressure sodium monochromatic yellow light. As the pressure increases to over one atmosphere, there is a broadening of the spectrum, giving a yellow and white light rather similar to an

High-pressure sodium lamps: quick data	
Colour rendering:	Poor to good (depends on chemical inside)
Luminous efficacy:	Good (80 to 140 lm/W)
Power range:	50 to 1000W
Life:	6000 to 24 000h
Power requirements:	AC from 100V upwards, needs control gear
Applications:	Industrial, commercial and street lighting

incandescent lamp. From start-up to 80% of rated light output takes about three minutes. After re-striking the same level is reached after only one minute.

Performance and life

Of the total input energy a high-pressure sodium lamp converts about 30% into visible radiation, 20% into infrared radiation, and 50% directly into heat. On average, high-pressure sodium lamps produce more than twice the lumen output of equivalent power mercury lamps, and yet have the same life (6000 to 24,000 hours). The luminous efficacy lies in the range 80 to 140 lm/W.

4.5 Recent Developments: Induction Lighting

Operating principles and construction

Induction lighting is a very recent and innovative development introduced originally by Philips Lighting, aimed at higher luminous efficacy and long life. Its operating principle is based on high-frequency induction.

The so-called QL induction lamp produces light using the same principles as fluorescent tubes, but uses a high-frequency antenna instead of gas discharge electrodes. The antenna works on the transformer principle: by applying a high-frequency alternating current to a primary coil wound onto a ferrite rod, an electromagnetic field is generated, which is coupled to the secondary coil formed by the gas filling the lamp (see Figure 4.17).

This is not dissimilar to the action of a microwave oven, but in the induction lamp the strong electromagnetic field ionizes the gas. Ultraviolet radiation is produced and converted to visible light in a similar manner to the reactions that take place in fluorescent tube. Indeed both types of lamp use a similar gas composition and a similar fluorescent powder coating on the inside of the lamp.

The light-producing process does not consume any material. Hence there is very little deterioration of performance with use. A built-in heat pipe removes all the heat generated by the antenna, so that no part of the lamp is hotter than 250°C, reducing the risk of breakage.

Performance and life

Typically an 85W induction lamp has a light output of 5500 lumens, which corresponds to a luminous efficacy of 65 lumens per watt. Its lifetime depends mainly on the electronic components of the control gear. It is estimated that more than 80% of induction lamps would survive after 60,000 hours continuous operation, and still provide more than 70% of their initial light output. This is some 60 times the normal lifetime of a tungsten filament bulb.

Other advantages of induction lamps are instant start-up, absence of flicker or stroboscopic effect, and low sensitivity of the light output to fluctuations in mains voltage.

The first induction lamps have recently become com-

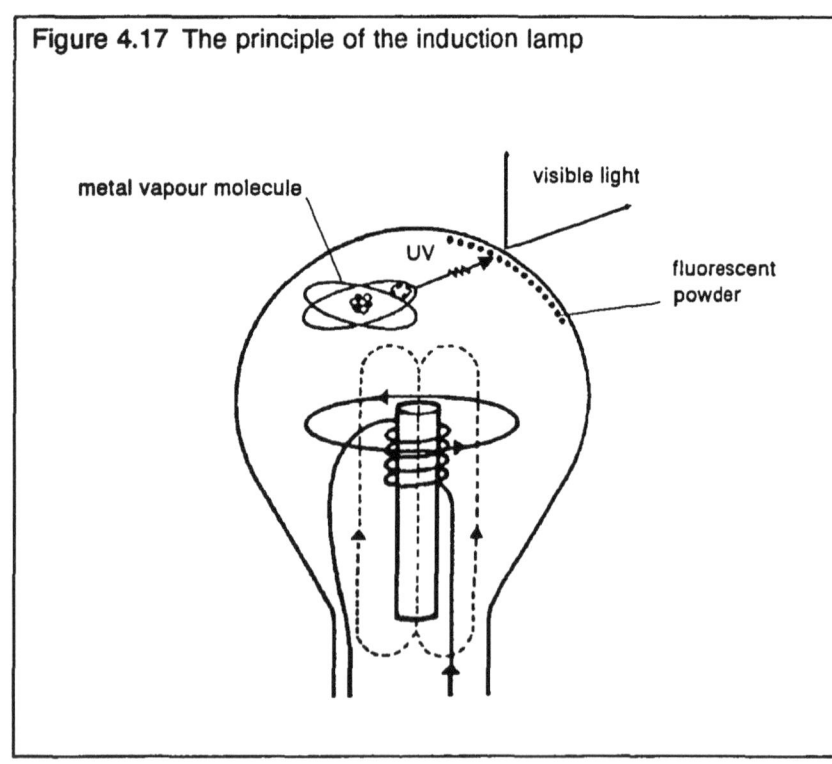

Figure 4.17 The principle of the induction lamp

metal vapour molecule

visible light

UV

fluorescent powder

mercially available. However, they are expected to be much more expensive than other types of lamp initially, and to be used primarily in places where it is particularly difficult and expensive to replace lamps. Their long lifetime justifies the extra cost.

4.6 Comparison of Electric Lamps

This chapter summarizes the main characteristics of a bewildering variety of electric lamps. Figure 4.18 indicates graphically the relative efficacies of the different types of lamps. Table 4.1 'General characteristics of lamps' and Table 4.2 'Applications of different lamps' summarize their principal characteristics, their primary advantages and disadvantages and the applications for which they are best suited.

The information contained in the tables are correct as of the date of publication. However due to rapid development in lighting systems, this information should be considered as indicative. In Figure 4.18 and Table 4.1, the luminous efficacy is given without control gear equipment unless it is incorporated in the lamp such as the CFL. Finally, the luminous efficacy is taken when the lamps are new.

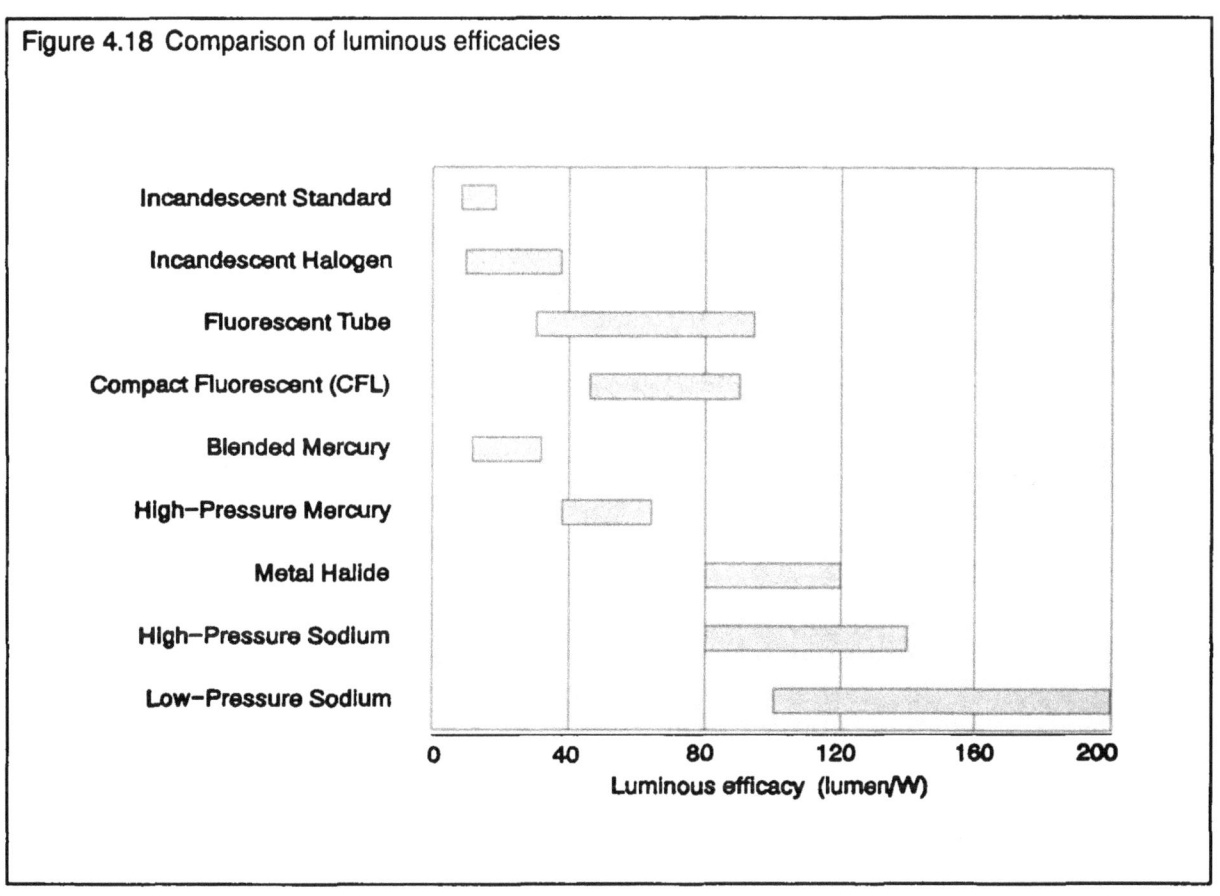

Figure 4.18 Comparison of luminous efficacies

Table 4.1 General characteristics of lamps

Type of Lamps	Average Luminous Efficacy (lm/W)	Luminous Efficacy Range (lm/W)	Sizes (W)	Correlated Colour Temperature (°K)	Colour Rendering (%)	Life (hrs)	Control gear needed	Warm-up time for 80% light output	Minimum starting temperature (°C)
Incandescent lamps									
standard	14	10/18	15/25/40/60/75/100/120/150 up to 1000	2600/2700	100	750/1000	NO	instant	any
miniature	10	8/12	0.75/1.2/1.8/2.7/3.6/4/5	2600/2700	100	15/200	NO	instant	any
incandescent halogen									
standard	20	17/24	75/100/200/300/500/750/1000/1500/2000	2800/3000	100	2000/4000	NO	instant	any
miniature	15	13/17	2/2.4/2.8/4/5	2800/3000	100	15/600	NO	instant	any
low-voltage (12/24 V)	20	12/30	5/10/15/20/35/50/65/75/100	3000/3100	100	2000	NO	instant	any
car headlight bulb (12V)	30	15/35	20/55/60/70/75/100	3000/3100	100	100/600	NO	instant	any
Low-pressure discharge lamps									
fluorescent lamps									
tube diam. 16mm	56	35/77	4/6/8/13	2700/5400	62/95	5000/8000	YES	1 to 2s.	any
tube diam. 26mm	75	50/100	16/18/32/36/50/58/60/70	2700/6500	50/95	5000/8000	YES	0.5 to 2s.	any
tube diam. 38mm	60	42/78	20/40/65/75/80/85/100/125	3000/4000	50/70	5000/8000	YES	1 to 2s.	any
compact fluorescent lamps (CFL)									
integral, inductive ctrl. gear	49	48/50	9/13/18/25	2700	85	8000/10000	NO	0.5 to 3mn	5
integral, electronic ctrl. gear	52	45/60	9/11/15/20/23	2700/4000	85	8000/10000	NO	< 15s.	any
modular (without ctrl. gear)	65	50/80	5/7/9/10/11/13/18/26 up to 55	2700/4000	85	8000/10000	YES	< 15s.	any
sodium low-pressure	150	100/200	18/26/35/55/90/135/180	Yellow monochromatic		6000/24000	YES	9 to 15mn.	-18
High-pressure discharge lamps									
mercury vapour lamps	50	38/63	50/80/125/250/400/700/1000	3400/4300	40/60	5000/24000	YES	5 to 7mn.	-20
blended lamps	25	15/35	100/160/250/500	3500	55	4000/6000	NO	instant	-18
metal halide lamps	100	80/120	70/150/250/400/1000/2000/3500	3000/6000	50/93	1500/15000	YES	3 to 5mn.	-18
high-pressure sodium lamps	110	80/140	50/70/100/150/250/400/1000	1900/2500	20/65	6000/24000	YES	3 to 5mn.	-40

Table 4.2 Applications of different lamps

Type of lamps	Typical applications	Colour rendering based on visual assessment
Incandescent lamps standard	General, localized or local lighting for homes, classrooms, offices, etc. To be avoided for long duration lighting due their high energy consumption	Good colour rendering but strongly emphasizes reds; yellows and greens to a lesser extent, blues strongly subdued
long life	Similar to standard incandescent, but more suitable when frequent lamp replacement presents difficulties. (i.e. typically such lamps last twice as long)	Similar to above
miniature	Torches, hand lamps, auxiliary lamps in cars, appliances, e.g. sewing machines, ovens	Similar to above
Incandescent halogen standard	Local and task lighting such as for desk office lamp, work spot light, torches In some cases, for security lighting system with infrared detector switch Can also be used for general lighting when purchase price of fluorescent lighting is not affordable	Good colour rendering but strongly emphasizes reds; yellows and greens to a lesser extent, blues strongly subdued
miniature	In torches as a replacement of standard bulb to increase their light output without increasing electricity consumption	Similar to above
dichroic (12/24 V)	Local lighting for health care (examination lamp, minor surgery, gynaecology, dental, intensive care, etc.)	Production of a white and cool light with up to 95% colour rendering
car headlight (12/24V)	Health care applications in emergency cases (preferably with white bulb)	Similar to incandescent lamps
Low-pressure discharge lamps fluorescent lamps (i.e. tubes)	General, localized or local lighting for homes, classrooms, offices, streets and sign illumination as a economical replacement of tungsten and standard halogen lamps. Should be avoided for short-duration lighting (e.g. toilets, cupboards)	The CCT depends on chemicals inside (warm, intermediate or cold) as does the colour rendering (CRI) Example of suitable CRI: 50 - 60 for sign, street lighting, 70 - 85 for offices, shops, hotels and factories engaged in critical work, CRI > 90 for clinical use
various sizes of tubes: tube diam. 16mm tube diam. 26mm	Portable lamps, desk lamps, built-in furniture lamps All applications. Should be preferred to 38mm as approximately 10% more efficient. High frequency 26mm tubes are even more efficient and can be dimmed	
tube diam. 38mm	Most commonly currently found, tends to be replaced by 26mm tubes	

Table 4.2   Applications of different lamps (cont.)

Type of lamps	Typical applications	Colour rendering based on visual assessment
Low-pressure discharge lamps (cont.)		
compact fluorescent lamps (CFLs)	General, localized or local lighting for homes, classrooms, offices, streets and sign illumination as an economical replacement of tungsten and standard halogen lamps	In general, provides both the warm colour appearance and the colour rendering of an incandescent lamp. Emphasizes oranges, greens, blues and violets. Subdues some yellows and deep reds
	Should be avoided for short-duration lighting (e.g. toilets) NEVER for dimmable illumination	
integral, inductive ctrl. gear integral, electronic ctrl. gear modular (without ctrl. gear)	Recommended when only few switchings on-off per day (e.g. two to six times) Allows more frequent switchings Recommended for very long period of illumination as the control gear lasts up to eight times longer than the tube	Similar to above Similar to above Similar to above
sodium low-pressure lamps	Out-door lighting: areas, streets, all-night security lighting PV powered street luminaires	Emits virtually only yellow light. All colours, except yellow, appear brown or black
High-pressure discharge lamps		
mercury vapour lamps	Industrial, commercial lighting such as for factories, public buildings, leisure premises Street lighting where a better colour rendering than that of a low-pressure lamp is required	Varies with chemicals used in the arc tube
blended lamps	Economically replaces incandescent lamps, especially where access for replacement is difficult NEVER for short-duration lighting	Similar to incandescent lamps but bluish
metal halide lamps	Demanding industrial, commercial lighting such as TV broadcasts, printing works, fashion stores	Depends on chemicals used in the arc tube (it is essential to consult a manufacturer)
high-pressure sodium lamps	General lighting in factories, warehouses, sport halls, public buildings Also for outdoor lighting	Emphasizes yellows, reds to a lesser extent Greens are acceptable, but blues are subdued

Enhancing Lighting Sources

5

5.1 Introduction

The light produced by a flame-based or an electrical system is valuable and should therefore not be wasted, but used to its fullest advantage. To use a light source efficiently will usually require some form of enhancement of the basic light source. The effectiveness of any lighting system depends on the following factors:

- Class of lighting system;

- Luminaire;

- Room surfaces.

These factors, if properly chosen, can greatly enhance a small light source. On the other hand, failure to consider either the properties of the room surfaces (e.g. wall colour) or failure to select properly either the appropriate luminaire or class of lighting will degrade the performance of even the best of light sources.

Daylight is a powerful source of light. During the day, daylight can enhance or completely replace flame-based or electrical light sources. Furthermore the effective use of natural light in a building is not only a feature which is highly regarded by most people, but can also permit great savings in energy consumption from artificial lighting. The last section of this chapter concerns using daylight to maximum effect.

5.2 Classes of Lighting

Three different classes of lighting system can be distinguished, according to their positioning relative to the task or object to be illuminated:

- General lighting;

- Localized lighting;

- Local lighting.

Table 5.1   Classes of lighting system

General lighting

- Almost uniform illuminance on working plane

- Regular layout of the luminaires

- Flexibility in task location

- Risk of low energy efficiency

Localized lighting

- Functional luminaire arrangement

- Required service illuminance on working plane

- Lower illuminance for other areas

- Lower energy consumption

Local lighting

- Illumination only over task area and immediate surrounding

- General lighting required (if affordable)

- Risk of glare for nearby workers.

- Economic supply of high illuminance level

General lighting systems provide an approximately uniform illuminance over the working plane, and the luminaires are usually arranged in a regular layout. In small rooms, and if only one luminaire is required, it will be installed in the centre of the ceiling. Such systems are generally energy-hungry and should be avoided whenever possible. For example, to provide illumination for corridors or lobbies and non-critical tasks, general lighting should be installed.

Localized lighting systems employ luminaires designed to provide the required service illuminance on work areas, together with a lower illuminance for other areas. Such systems consume less energy than general lighting systems.

Local lighting provides illumination only over the small area occupied by the task and its immediate surroundings. Local lighting must be positioned to minimize shadows, veiling reflections and glare. Most local lighting systems are accessible and adjustable. This increases wear and maintenance costs, but provides precise control. Local lighting can be the most energy-efficient method for providing task illumination.

Both local and localized lighting offer scope for switch control of individual luminaires, which can be turned on and off according to individual needs. The principal characteristics of the lighting classes are summarized in Table 5.1.

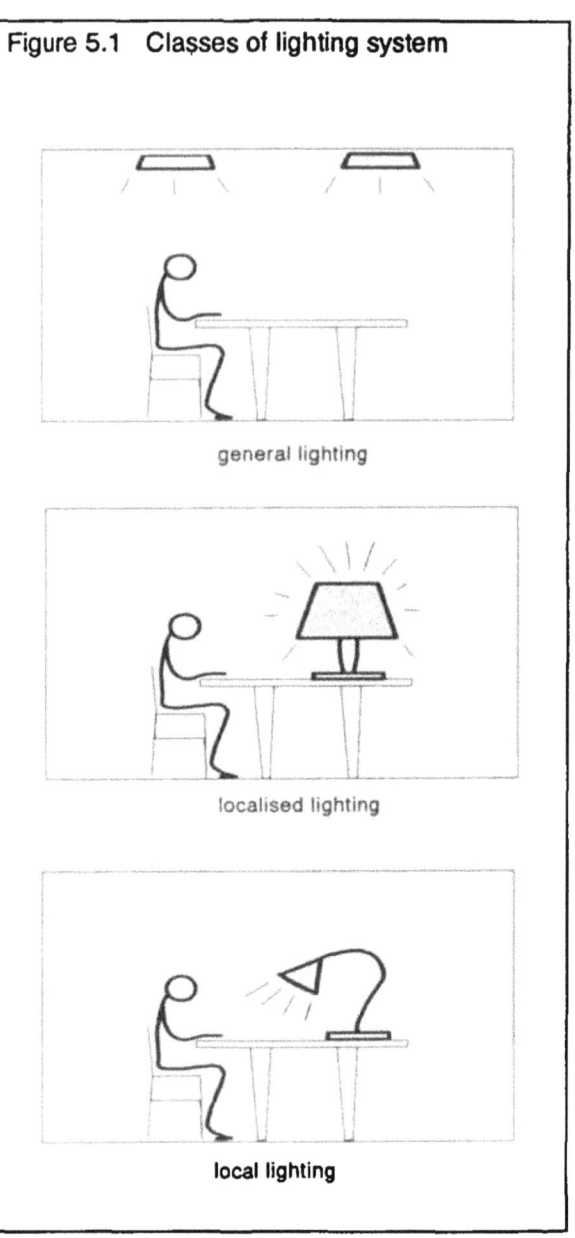

Figure 5.1 Classes of lighting system

general lighting

localised lighting

local lighting

5.3 Luminaires

Bare lamps (bulbs) are seldom used in lighting installations as they often cause glare, do not distribute the light in an efficient and effective manner and may look unattractive. Therefore, either for aesthetic or technical reasons, lamps are usually placed into translucent or reflective 'shades', which are intended to reduce (or redistribute) their illuminance to required levels, prevent glare, provide mechanical protection and improve their general appearance. Such devices are called luminaires, or fixtures and fittings.

By definition, a luminaire is a complete lighting unit comprising one or several lamps, which has the following functions, namely to:

● Control and distribute the light;

● Support and protect the lamps mechanically;

● Electrically connect, operate and control the lamps;

● Maintain the lamps at the correct temperature.

Many countries have established standards, which specify requirements for luminaire construction. These generally relate to such factors as protection against electric shock, ingress of solid bodies (dust) and moisture.

They also define the type of supporting surface for which luminaires are designed. The lighting designer should generally take care to choose luminaires, which comply with standards in the country concerned.

Construction materials

The design of luminaires includes features to comply with different needs. For example, it may be necessary to concentrate (focus) the light, or to diffuse (spread) it. This is achieved by using the reflection, transmission and refraction properties of different materials shaped into different forms.

The most commonly used materials are:

- Specular or diffuse aluminium;

- White synthetic or porcelain enamelled steel;

- Clear, opal, diffuse or coloured glass;

- Different varieties of clear or white plastic materials: e.g. acrylic, polystyrene, vinyl, etc.

Reflective materials can be shaped into plane, spherical, elliptical or parabolic surfaces, depending on the desired shape and direction of the beam. Clear or translucent materials can be shaped into prismatic forms or lenses.

Luminaire classification

The CIE (Commission Internationale de l'Eclairage) classifies luminaires into six types according to the percentages of total light output emitted above and below the horizontal:

- *Direct lighting*: when luminaires direct 90 to 100% of their output downward;

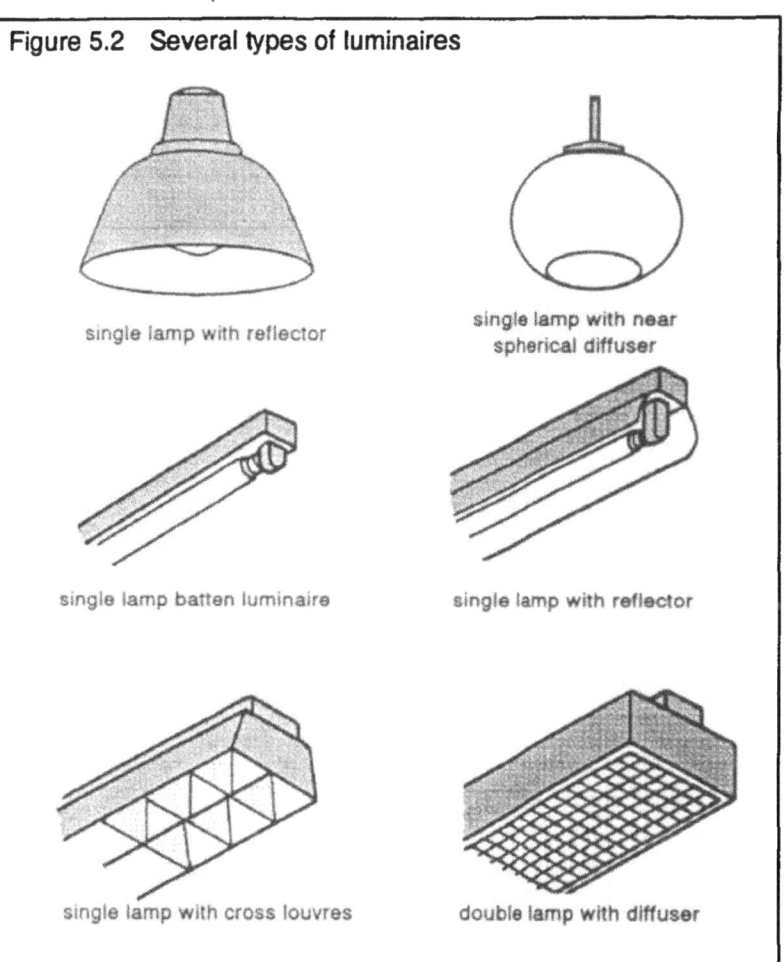

Figure 5.2 Several types of luminaires

single lamp with reflector

single lamp with near spherical diffuser

single lamp batten luminaire

single lamp with reflector

single lamp with cross louvres

double lamp with diffuser

- *Semi-direct lighting*: when the distribution of light output is predominantly downward (60 to 90%), but with a small upward component;

- *General diffuse lighting*: when downward and upward components are about equal (each 40 to 60% of luminaire total output);

- *Direct-indirect lighting*: when downward and upward components are about equal (each 40 to 60% of luminaire total output), but with very little light being emitted at angles near the horizontal;

- *Semi-indirect lighting*: when the distribution of light output is predominantly upward (60 to 90%), but with a small downward component;

- *Indirect lighting*: when luminaires direct 90 to 100% of their output upward to the ceiling and upper side walls.

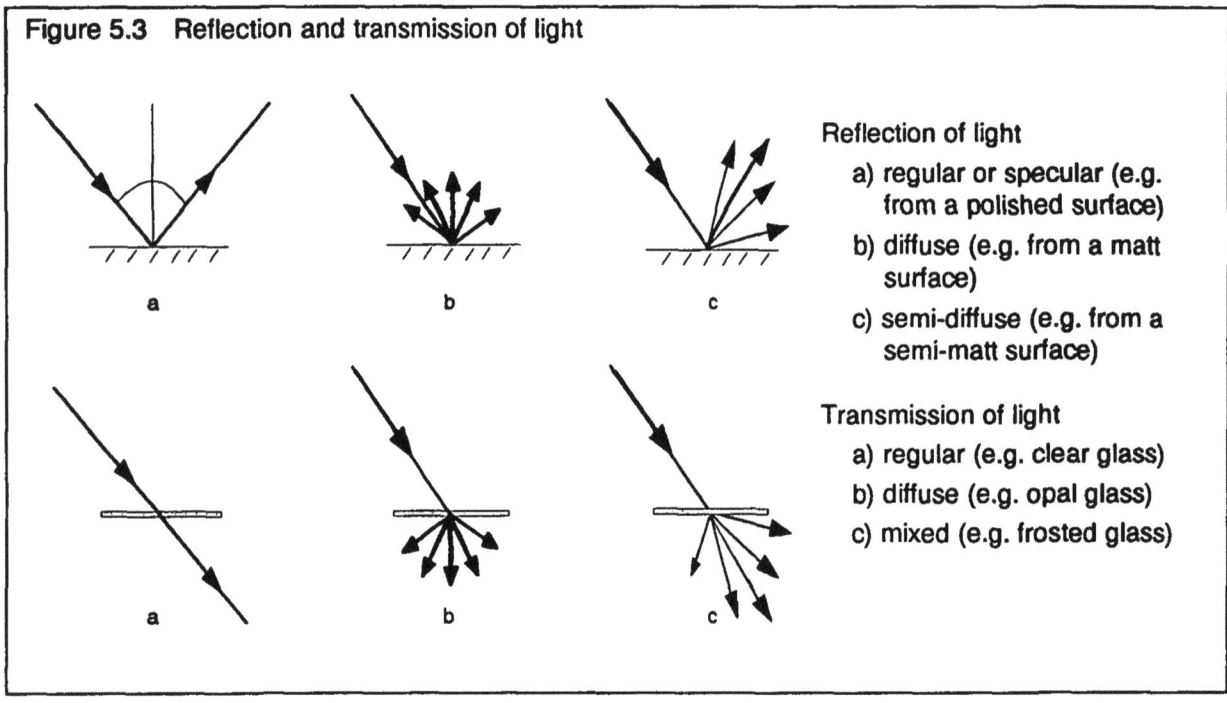

Figure 5.3   Reflection and transmission of light

Reflection of light
  a) regular or specular (e.g. from a polished surface)
  b) diffuse (e.g. from a matt surface)
  c) semi-diffuse (e.g. from a semi-matt surface)

Transmission of light
  a) regular (e.g. clear glass)
  b) diffuse (e.g. opal glass)
  c) mixed (e.g. frosted glass)

Luminaire light control

Several methods are used to control luminaire light output so as to eliminate glare and distribute the light efficiently and effectively, namely diffusers, baffles (or louvres) and reflectors.

Diffusers

Diffusers are used on the sides and backs of luminaires in order to redirect or scatter light and eventually to reduce luminances in the glare zone (45° to 90°). They are made either of glass or plastic materials with high transmittances and shaped into different forms:

Table 5.2	Transmittance values of several diffuser materials for visible light
Material	(%)
No glass	100
Clear glass	90
Depolished glass	55 - 65
Opalescent glass	60 - 85
Silk lamp shade	5 - 55
Parchment paper	40 - 45

- Clear prismatic lenses are used in different patterns (series of conical or pyramidal-shaped prisms) to obtain the desired intensity distributions

- Translucent sheets are used to scatter the light in all directions rather than directing it to particular zones

- Combinations of prismatic lenses and flat translucent sheets can be used in order to obtain particular distributions of luminous intensity

Baffles and louvres

A single shielding element placed between lamps in a two-or-more-lamp luminaire is referred to as a baffle. Baffles are usually V-shaped. A louvre is a group of baffles arranged as in an egg-crate. Louvres may be straight or parabolic. Also, baffles and louvres are characterized by their shielding angle, which is the angle between the horizontal and the line of sight above which all objects are concealed.

When low ceiling brightness is required, louvres have small hexagonal or square cells with a 45° shielding angle, made of specular, translucent or opaque materials. These constructions result in low luminaire efficiencies, often below 50%.

Reflectors

Reflectors are mounted above lamps to redirect the upward component of luminous flux downwards. They come in two types: specular (or semi-specular), and diffuse.

Specular and semi-specular reflectors are generally used with incandescent and HID lamps. Etched, polished, brushed, plated or anodized aluminium is normally used to produce their reflecting surfaces, which may have circular, parabolic, elliptical or combined contours. Self-adhesive aluminium foil has become available to improve the reflectance of existing fluorescent light luminaires.

Reflectors with a high reflection factor can almost double the light output in some cases. For comparison, Table 5.3 gives reflection factors for reflector materials as well as for other building materials.

The use of reflectors may make the other components of the luminaire operate at a higher temperature, since they reflect heat as well as light, which may result in reduced output and lifetime for the lamps, and thermal stress and failure for the diffusers. Reflectors can be locally made to enhance or to modify the light output of existing luminaires (see Figure 5.4).

Table 5.3	Reflection factor of different materials for visible light
Mirror	0.80 - 0.90
Enamel (white)	0.65 - 0.75
Aluminium (polished)	0.65 - 0.75
Aluminium (matt)	0.55 - 0.60
Chrome (polished)	0.60 - 0.70
Tin plate	0.68 - 0.70
White plaster work (new)	0.70 - 0.80
White plaster work (old)	0.30 - 0.60
White distemper	0.65 - 0.75
White oil paint	0.75 - 0.85
Concrete (new)	0.40 - 0.50
Concrete (old)	0.05 - 0.15
Yellows curtains	0.30 - 0.45
Red, blue or dark brown curtains	0.10 - 0.20

In Zaire, in 1986, as part of the largest Health/Solar development programme to date. 8W fluorescent luminaires with diffusers were supplied for 750 lighting systems for health centres and hospitals. These luminaires were not designed for health needs but for the leisure market. As a result, they were fitted on-site with a metallic reflector and support structure to make desk lamps for nurses and doctors offices. Projectors for deliveries and surgical operations were made of two to three luminaires mounted together and backed by a metallic reflector, hanging on a mast.

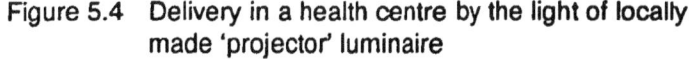

Figure 5.4 Delivery in a health centre by the light of locally made 'projector' luminaire

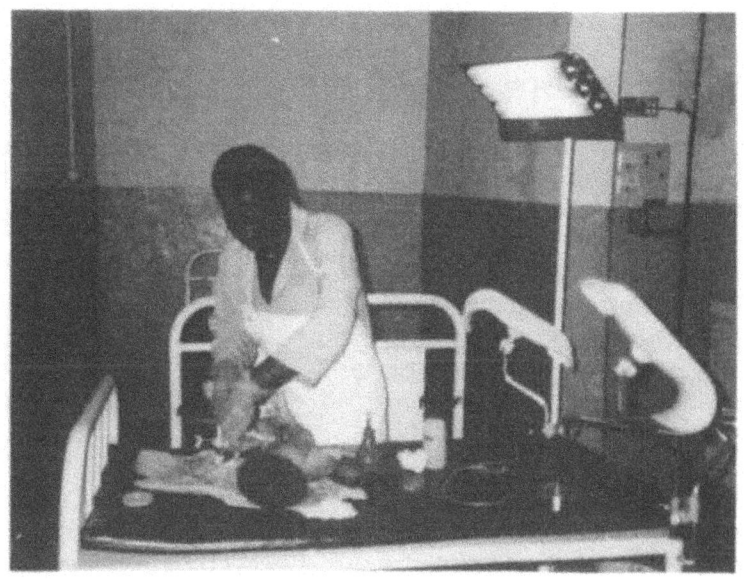

Luminaire characteristics

Luminaires can be characterized according to various different criteria. The most important of these are:

- *Mounting position*: luminaires are usually either recessed into the ceiling, surface mounted or hanging;

- *Luminaire efficiency*: is the ratio of the lumens emitted by the luminaire to those emitted by the lamps. Efficiencies lower than 70% indicate poor luminaire design;

- *Utilization factor*: is the luminous flux which reaches the working plane as a ratio of the luminous flux emitted by the lamps. It depends on luminaire efficiency, but also on the distribution and spacing of luminaires, room proportions and the reflectance of the room surfaces;

- *Glare index*: varies with luminous intensity distribution, room proportions, room surface reflectances and the room and luminaire orientation relative to the line of sight.

These characteristics are generally provided by manufacturers for each type of luminaire they produce. It is recommended to refer to manufacturers' specifications as part of the design process for lighting installations.

5.4 Room Surfaces

The properties of room surfaces, i.e. reflectance and colour, can strongly influence the effectiveness of a lighting installation in an interior. For this reason, the room properties should be assessed and, if necessary, modified to enhance the light source.

The reflectance of surfaces is the ratio of the luminous flux reflected from the surface to the luminous flux incident upon it. The value is always less than one and is expressed as either a decimal or a percentage. Table 5.4 gives some average reflectance values for visible light which can be used in lighting design (see Chapter 9).

For interiors lit from the ceiling as is often the case, the significance of the ceiling reflectance increases as the room area increases. Consequently in a large room, the ceiling should be painted in white. Where indirect lighting is used, the ceiling should be white regardless of the room size.

In small rooms, it is the wall reflectance which may be the more influential. Furthermore, high wall reflectance usually enhances the illuminance on the working plane and increases the uniformity. Dark floors will tend to make ceilings and walls look underlit especially when daylight from windows is used. On the other hand, light floors require frequent cleaning.

It is often not practical to alter the reflectance of the floor because durable good quality floor paint or tiles are expensive. However, it should be realized that painting ceilings and floors in white or light-coloured paint or at least whitewash, is likely to be cheaper than installing a more powerful lighting system.

The colour of surfaces and even objects in a room can create a variety of atmospheres (e.g.

Table 5.4 Average reflectance values for visible light (%)		
Colours	Ceiling	Wall
White or light ivory, light blue, clean	70	50
Same colour as above but dirty	50	30
Other	10	10

An average reflectance value of 30% can be taken for windows and 20% for floors.

'warm', 'restful' or 'active'). There are some rules of colour harmony which have little basis in fact, e.g. a red colour may not have the same meaning in a western culture that it does in others. As several colours of paints e.g. light ivory or very light blue may have the same reflectance (it makes no difference for the lighting system sizing) and as lighting users may feel more comfortable in one of these colours, it is always advisable to involve the light users in the process of selecting colours.

5.5 Daylight Contribution Enhancement and Assessment

This section aims to give hints for using daylight effectively, as well as to evaluate the daylighting contribution.

The penetration of direct sunlight into a building creates a more pleasant living environment and gives a sense of the passage time throughout the day. However placing openings (windows or skylights) in a building's exterior presents not only many advantages, but also several disadvantages, which are summarized in Table 5.5 'Advantages and disadvantages of daylighting'.

Getting the best from daylight

In daylight design, which ideally starts at the architectural conception of a building, many choices have to be made in the proper order, as follows:

- Identify the directions with the most interesting and pleasant views;

- Consider sunlight availability on the site. Select appropriate orientation for the building and its facades. Remember that it can be more difficult to achieve good shading for west and east facades than it is for south-facing orientations;

- Organize the indoor spaces with respect to daylighting requirements. Do not hesitate to create apertures in the roof as a sky light source is more economical than lateral openings (the glass surface can be reduced by up to five times for an equal intensity of light);

Table 5.5 Advantages and disadvantages of daylighting

Advantages

- reduces artificial lighting energy consumption
- provides ventilation
- allows indoor green plant growth
- gives a feeling of open place
- gives visual and psychological relief and well-being
- Provides additional emergency exits
- gives a heat gain in winter in temperate climate.

Disadvantages

- is a potential source of glare
- can cause overheating in summer or hot seasons
- is uncontrollable and constantly varying
- transmits noise
- increases maintenance costs for cleaning and breakage
- increases construction costs
- may compromise security: easily penetrated
- transmits dust and other pollutants (if windows/apertures are open).

- Distribute and size apertures with respect to volume of the indoor spaces to be lit;

- Design solar shading devices (e.g. curtains, blinds or overhangs);

- Particular attention should be given to ensuring low or easy maintenance;

- Remember that rooms which receive daylight in small quantities often need warmer colours to improve the luminous environment;

- Design a sound artificial lighting system which takes full account of daylit zones. This involves taking care of the location of the luminaires and the number of lamps connected to each light switch. Evaluate the possible benefits of task lighting as well as automatic control of overall electric lighting with respect to daylight availability.

When installing windows or organizing the activities within a room, each opening should be carefully designed and evaluated in order to optimize its lighting contribution and, at the same time, minimize the negative environmental factors (e.g. overheating). The following rules should be considered:

- Avoid direct sunlight penetration by correct solar protection;

- Provide enough openings on opposite walls to improve uniformity and reduce shadow effect;

- Use clear glazing material (if any);

- Use high windows to let light reach the inner portion of the room.

When planning to make the best use of daylight, it should be kept in mind that natural light is not always free of cost. Indeed, the provision of a well-designed window may well cost much more than the portion of plain wall it replaces.

Assessment of daylight contribution

The extent to which daylight contributes to the illumination of an interior is best described by a parameter called the Average Daylight Factor (ADF). There are several methods to calculate the average daylight factor. Appendix C 'Calculation of the Daylight Factor' gives one simple method for calculating the ADF, as well as an example calculation.

It has been found that when the average daylight factor (calculated with the method given in Appendix C) is 5% or more, an interior will generally look well daylit. When the average daylight factor is less than 2%, the interior will not be perceived as well day-lit and electric lighting may be in constant use during the day. When the average daylight is between 2% and 5%, the electric lighting should be planned to take full advantage of available daylight. Localised or local lighting may be particularly advantageous, using daylight to provide the general surround lighting. Table 5.6 shows the approximate relationship between daylighting and electric lighting based on the values of the average daylight factor as calculated with method described in Appendix C.

Table 5.6 Influence of average daylight factor values on electric lighting
ADF = Average daylight factor
ADF < 2%
Electrical lighting in use throughout the day
2% < ADF < 5%
Localized or local electric lighting with daylight for general lighting
ADF > 5%
Minimal or no use of electric lighting during the day

Energy for Flame-based Lighting 6

6.1 Introduction

This chapter is about the fuels that are commonly used to produce light by direct combustion. Most of these fuels are fossil fuels such as natural gas and petroleum derived from solar energy many thousands of years ago and stored under the earth's surface. Unfortunately many developing countries do not have indigenous fossil fuel resources and have to import them using scarce foreign currency.

Some fuels such as vegetable oils, biogas and alcohols are produced from biomass. The energy in biomass is derived from the daily sunshine which is captured by the plants by photosynthesis. Therefore biomass fuels are renewable energy fuels and are often called biofuels. Biofuels are likely to play an important future role on a long-term basis, when fossil fuels become more scarce.

The fuels described in this chapter are:
- Natural oils and waxes;
- Kerosene;
- LPG bottled gas (butane/propane);
- Alcohols;
- Natural gas;
- Biogas;
- Calcium carbide.

The last section summarizes the characteristics of various fuels for flame-based lighting, including, for comparison, fuels used for other applications.

6.2 Natural Oils and Waxes

Since the invention of the hollow-stone oil lamp in prehistoric times, up to the discovery of crude oil resources in Pennsylvania in the middle of the nineteenth century, the only fuels used in oil lamps were natural oils, most often of vegetable origin. Olive oil, cotton-seed oil, sunflower-seed oil, rape-seed or palm oil have all been used depending on the location.

Natural oils continued to be used for lighting in Africa and Asia long after the beginning of this century, but now their use is limited only to the very poor sections of the community in rural and some urban areas. Elsewhere they have generally been replaced by kerosene to fuel hurricane lanterns.

Wax and tallow were used to manufacture the early candles. Wax was obtained from bees' honeycomb and tallow from mutton, pork or beef fat, which people carefully saved from

---

Energy content of fuels: heating value

Each fuel differs from another by the amount of energy it holds per unit of weight. The total energy content is characterized by either its net or gross heating value. The *gross heating value* equals the heat released by complete combustion under standard conditions, including the latent heat of the combustion gas (e.g. water condensate). The *net heating value* does not take into account the latent heat of gas, therefore it is less than the gross heating value. It is sometimes referred to as the *lower heating value*. For instance, the gross heat value of kerosene is equal to 48MJ/kg, while the net heating value is 45MJ/kg. Usually, the heating value referred to is the net heating value.

The importance of net heating value can be seen through the following example. One kilogram of kerosene (i.e. 1.25 l) burning in a 25% efficient cookstove will heat 35 l of water from 25°C to 100°C. Assuming the same efficiency for a woodstove, one kilogram of wood, which has a smaller net heating value of 15MJ/kg, will heat only 12 l of water from 25°C to 100°C.

their kitchen wastes. Tallow candles were home-made products. Present day candles are made of paraffin wax obtained from crude petroleum oil on a large industrial scale. The heating value of petroleum wax is about 36MJ/kg.

6.3 Kerosene

The fuel oils commonly called kerosene (or sometimes paraffin) include three different types of fuels: illuminating oil, vaporizing oil and long burning oil. Kerosene is obtained from crude petroleum by refining (e.g. distillation).

For illuminating purposes, oils consisting mainly of paraffins (illuminating oils) give less smoke, and a larger, whiter and more luminous flame than the others. Oil composition is less important when it is burned in lamps using incandescent mantles. An important characteristic of illuminating oil is its 'smoke point', which is the flame height in millimetres at which smoking commences. It should be at least 25mm. The net heating value of kerosene is about 36MJ/litre.

Kerosene is the most important fuel used for lighting in developing countries. India alone is reported to use around 3.6 million tonnes per year of kerosene for lighting in more than 100 million hurricane lanterns. In Mali, a typical medium-sized African country with 9.5 million inhabitants, the annual consumption is around 10,000 tonnes, and kerosene is sold at petrol stations at the equivalent of US$0.75/litre. In Mali, few people buy kerosene at petrol stations, because most rural people live too far from the main roads. Usually, kerosene is made available in remoter villages by local merchants, who sell it in very small quantities on a daily basis. In Mali, the daily cost of kerosene for the average family for lighting is estimated at US$0.17 per day per lamp. Although this arrangement is consistent with rural peoples'

purchasing power, it makes the real price of kerosene very high, approaching double the petrol station or oil company official price.

6.4 LPG Bottled Gas (Butane/Propane)

LPG bottled gas is made of butane and propane which are hydrocarbon gases produced during the refining of petroleum. They are gaseous at normal temperatures but can easily be compressed so as to become liquid under ambient temperature conditions.

When these gases are liquefied under pressure, they are commonly known as Liquefied Petroleum Gases (LPG). Butane and propane burn with a bluish flame and can only be used for lighting in conjunction with a rare-earth incandescent mantle. One kilogram of liquefied propane when released at normal temperature (20°C) and atmospheric pressure will give more than $0.53m^3$ of gas, and one kilogram of butane will give nearly $0.44m^3$ of gas. The net heating value of LPG can be taken at 45MJ/kg which is equivalent to the net heating value of kerosene.

Commercial LPG is often a mixture of butane and propane. It is transported from refineries in special tanker vehicles to gas filling

Figure 6.1 Kerosene and gasoline rural filling station in Indonesia. Note: The brand name of the gas is SOLAR but unfortunately these fuels are not biofuels!

stations, where it is transferred into transportable refillable cylinders or bottles capable of holding 2.7kg, 6kg, 12kg or 16kg up to 47.2 kg of gas.

LPG (usually butane) is also available in disposable (and expensive) throw-away cartridges, which contain about 0.2kg of gas each. They are for use with camping lamps and cookers. It is especially expensive in this form and tends to be aimed primarily at the leisure market for the wealthy from the cities, although the disposable containers are retailed in certain rural areas.

Safety is of critical importance in the handling of LPG as there is considerable danger of fire or explosion if leakage occurs. Any source of ignition and any possibility of gas escaping and mixing with air should be avoided. LPG cylinders should always be in good condition (without dents or visible damage) and they should never be stored in direct sunlight or near a source of heat.

The price of gas is strongly influenced by transportation costs. For example, a kilogram of LPG costs US$0.37 in Dakar, Senegal; it costs US$1.10 in Bamako, Mali, but is sold at US$0.83 since European Commission subsidies have been introduced in all the nine Sahelian countries of West Africa in an attempt to get LPG substituted for firewood or charcoal for cooking, and hence to reduce the desertification taking place as a result of deforestation.

Transportation costs of cylinders to and from gas refilling stations limit the use of gas to areas close to these stations. Since there are very few stations in each country, LPG gas is only available to relatively few people.

6.5 Alcohols

Diminishing world reserves of petroleum together with the heavy burden of petroleum-product importation costs on the economies of many of the poorer countries, generate an urgent need to find a renewable energy replacement for liquid fuels. The production of alcohols (e.g. methanol and ethanol) that can be derived from crops such as sugar cane or from cereal crops is one possibility. While viable methanol production is still at the

research stage, ethanol has been massively produced by fermentation in countries such as Brazil and the Philippines. The heat value of ethanol is 21.9MJ/litre against 36MJ/litre for kerosene.

Ethanol can efficiently replace kerosene in incandescent mantle pressure lanterns. Ethanol is not suitable for the basic wick type of lamps as it has an almost non-luminous blue flame. An efficient pressure lantern has been developed in India which can burn ethanol and has a light output equivalent to a 100W electric incandescent bulb. (For further information see bibliography, Anil K. Rajvanshi.)

Although traditionally ethanol has been produced mainly as a by-product of the sugar industry, it can also be obtained from cassava and from sweet sorghum. The latter may offer a means of producing cheap alcohol: one hectare of sweet sorghum reportedly yields 3000 to 4000 litres of alcohol per annum which can be produced in decentralized mini distilleries of 5000 to 10,000 litres/day.

A snag inhibiting the use of alcohol for lighting is that at the present time it is often more expensive than kerosene and less easily procured in remote rural areas. Finally, the production of ethanol can hardly be justified only for lighting needs and should be part of an integrated energy strategy (e.g. energy for transport, cooking, lighting, etc.).

6.6 Natural Gas

Natural gas consists primarily of methane (CH_4), the lightest of the hydrocarbons. Natural gas is produced directly from the gas wells drilled by the multinational oil companies. Large deposits of natural gas have been found in the USA and Canada, the former Soviet Union, the Middle East, North Africa, the North Sea, Bangladesh, China, Indonesia, etc. Natural gas is often associated with oil but sometimes occurs separately.

Raw natural gas emerging from a gas well requires only purification to ensure safe distribution by means of pipelines. Natural gas can also be transported in bulk in the form of Liquefied Natural Gas (LNG) which is obtained by cooling the gas to -160°C under normal pressure (600 m³ of gas will condense down to only 1m³ when lique-fied).

When burning, natural gas produces a non-luminous flame, hence it is necessary to use an incandescent rare-earth mantle to obtain visible light with a relatively good efficiency for a flame-based lighting system. The net heating value of natural gas depends on its content of methane. On average, it is around 35MJ/m³ (at 20°C and atmospheric pressure). Natural gas used to be used for lighting, but nowadays is mainly burnt for heating, cooking and in electrical power stations.

Figure 6.2   Domestic biogas floating drum unit in the Philippines

6.7 Biogas

Biogas is a flammable gas produced by microbes when organic materials (such as sewage, animal and human manure, or food-processing wastes) are fermented within a certain range of temperature and moisture content and under anaerobic (oxygen-free) conditions. This is called anaerobic digestion. In ponds and marshes where there is a high content of rotting organic material, bubbles of biogas can often be seen rising to the surface.

Biogas consists of 60 to 70% methane (CH_4) (which is chemically identical to natural gas) and 20 to 25% of carbon dioxide (CO_2). The remainder is composed of small quantities of hydrogen sulphide (H_2S which has a distinctive 'rotten egg' smell), carbon monoxide and several other hydrocarbons. CO_2 is inert and tends to dilute the net heating value of the biogas compared with natural gas.

The net heating value of raw biogas is 25MJ/m³ (at 20°C and atmospheric pressure). In other words, one cubic metre of biogas has the same energy content as 0.7 litre of gaso-

line. When burning, biogas produces a non-luminous blue flame, hence it is necessary to use an incandescent rare-earth mantle to obtain visible light with a reasonable efficiency for flame-based lighting using biogas.

Animal dung is the easiest feedstock to use for a biogas plant. Raw vegetable matter usually needs to be treated before it can be used: it can be chopped up or minced, or composted for a few days before being added to the biogas plant. The volume of biogas produced depends on the feedstock, which is

Table 6.1	Biogas yield from various feedstock			
	Total Waste volume (l/day)	Total dry Waste solid (kg/day)	Biogas production (m³/day)	Energy yield (MJ/day)
Cow	36.0	4.7	0.40	9.90
Pig	14.0	0.7	0.30	7.50
Hens (x100)	13.0	2.2	1.00	24.90
Human	1.4	0.1	0.05	1.25

Note: The figures given above are only indicative.

added to the plant. Average production rates of dung per single typical animal are given in Table 6.1.

Various designs of biogas digesters are available. Only three types are commonly used on a small scale in rural areas: the floating drum ('Indian digester'), the fixed dome ('Chinese digester') and flexible bag digesters. Biogas production depends on many factors: the design and volume of the digester; on the temperature of the feedstock (ideally between 30 to 35°C); retention time (20 to 40 days) and loading rate and feedstock carbon/nitrogen ratio.

Lighting with biogas is practised in parts of India and China, the two main countries to have major programmes to popularize the production and use of biogas from waste materials. It is unlikely that it will generally be worth producing biogas primarily for lighting, but if it is available for cooking, a small proportion can readily be used for lighting. Larger farm-scale biogas units can also be used to power internal combustion engines that in turn can generate electricity for lighting.

6.8 Calcium Carbide

Calcium carbide is a hard, rock-like material that reacts immediately on contact with water to produce acetylene gas. Acetylene is highly inflammable and burns in air with a bright flame. Calcium carbide is produced mainly to provide an economical supply of acetylene for the metal work industry (e.g. flame-cutting and welding) but can also be used for lighting applications.

The basic raw materials for producing calcium carbide are limestone ($CaCO_3$) and coal or coke. Calcium carbide is prepared by heating quicklime (produced by heating limestone) and carbon between 2000 and 2200°C in an electric arc furnace. The equation is as follows:

Figure 6.3 Large-scale biogas plant. In this case, the biogas can be burnt in biogas lamps to provide lighting, but could also power internal combustion engines that in turn can generate electricity for lighting with a better energy efficiency

$$CaO + 3C > CaC_2 + CO$$

quicklime + coke > calcium carbide + carbon monoxide

On average, 0.85kg of quicklime, 0.5kg of coke and 3kWh of energy are necessary to produce 1kg of calcium carbide. The manufacture of calcium carbide is an energy-intensive process which does not lend itself to small-scale applications. Typical calcium carbide furnaces produce in the range of 120 to 500 tonnes per day requiring an electrical power of 20MW and 70MW respectively. Molten calcium carbide is poured into moulds to solidify. It is then crushed to produce the required particle sizes. For carbide lamps, carbide is usually required to be in graded lumps of 15 to 25mm.

Care is needed with the storage and handling of calcium carbide, since acetylene can easily be produced from no more than prolonged contact with moisture in the air. Moreover, acetylene is capable of causing a violent explosion in a wide range of mixtures with air as well as being capable of exploding without air by spontaneously breaking down into carbon and hydrogen, as it is potentially unstable. Hence it is not normally stored as a gas except in special cylinders in which it is dissolved in acetone in order to avoid the risk of spontaneous explosive decomposition of the gas.

The total energy input necessary to produce 1kg of calcium carbide is approximately 8.3kWh including the calorific value of coke and the electricity input (0.5kg coke x 10.6kWh + 3kWh electricity). On the other hand, the energy output per kg of calcium carbide is 4.5kWh (0.35kg x 46.2MJ/kg (calorific value of acetylene) = 16MJ = 4.5kWh). Therefore the energy conversion efficiency is 54%. Hence, calcium carbide is a reasonably energy-efficient way of storing energy (e.g. electricity produced by a large hydro-power plant).

If the fuel is well packaged it is relatively safe and portable. Therefore this form of energy for lighting may merit more serious consideration than it gets at present, especially for countries having plentiful low-cost hydro-electric power combined with a source of limestone and carbon (possibly charcoal) for the production of calcium carbide.

Unfortunately, few developing countries can consider local production of calcium carbide and therefore they would be dependent on imports. Importation prices are quite low (US$50 to $70 for a 100kg drum of 15 to 25mm grade particles), but the cost of the carbide to the users is considerably augmented by packaging and transport costs (from US$1 to $2 per kg). Today, imported carbide is used in many developing countries as a means to generate acetylene for flame-cutting and gas-welding in workshops. Special gas generators are available for this purpose or in some cases (e.g. in Zaire) are locally made from empty 35kg gas bottles.

6.9 Characteristics of Various Fuels

Other fuels than those for flame-based lighting are included for comparison and also because some of these fuels can be used for electricity generation for lighting purposes. The prices given are approximate.

Table 6.2   Main characteristics of various fuels

Type of Fuel	Net heating Value (per mass)	Density (at 1atm. 20°C)	Net heating Value (per volume)	Price range	Fuel for flame-based lighting	Fuel for electrical lighting
Solid	MJ/kg	kg/l				
Dry wood	15.0	0.5-0.8	n/a	0 to 0.1 $/kg	YES	YES
Wax	36.0	0.90	n/a	1.0 to 3.0 $/kg	YES	NO
Carbide	20.6	2.34	n/a	1.0 to 2.0 $/kg	YES	NO
Liquid	MJ/kg	kg/l	MJ/l			
Gasoline	44.0	0.74	32.6	0.4 to 1.2 $/l	NO	YES
Kerosene	45.0	0.80	36.0	0.5 to 1.5 $/l	YES	NO
Diesel/gas oil	42.8	0.84	36.0	0.1 to 1.0 $/l	NO	YES
LPG	45.0	0.54	24.3	0.3 to 1.5 $/kg	YES	YES
Ethanol	27.6	0.79	21.9	0.7 to 1.5 $/l	YES	YES
Gas	MJ/kg	kg/m³	MJ/m³			
Natural gas	46.4	0.75	34.8	0.1 to 0.5 $/m³	YES	YES
Propane	42.5	2.02	85.8	0.2 to 0.8 $/m³	YES	YES
Butane	43.3	2.58	111.8	0.2 to 0.8 $/m³	YES	YES
Methane	46.7	0.72	33.5	0.1 to 0.5 $/m³	YES	YES
Biogas	27.7	0.90	24.9	variable $/m³	YES	YES
Acetylene	46.2	1.17	54.0	2 to 4 $/m³	YES	NO

Electrical Storage: Batteries

7

7.1 Introduction

Where possible, it is best to use electrical lighting, since this offers unrivalled efficiency, light output, controllability and absence of such undesirable features as high heat output, smoke and fumes, fire hazard or noise. Furthermore, it is what the majority of people everywhere expect and aspire to as we near the end of the twentieth century.

Since most small-scale methods of electricity generation are only intermittently available, some form of electricity storage or battery is generally an essential component, if lighting is to be available at the 'flick of a switch' at any time.

In the field, batteries are very often the least reliable component in a lighting system for many reasons, including poor battery quality and lack of, or inadequate, maintenance. The aim of this chapter is to provide information and guidelines to reduce the potential for battery problems as far as possible. The central point is that there is no such thing as a universal battery: i.e. a single type of battery cannot cover all applications.

This chapter reviews the state-of-the-art of the various types of batteries (or cells) that are presently available and reviews their advantages and disadvantages. Some simple rules for selection, choice, sizing, installation and maintenance are also given. Towards the end of the chapter, typical ways of transforming the stored DC electricity into AC current are also given.

Batteries can be sub-divided into the following types:

● Primary cells or dry batteries;
    -standard zinc-carbon
    -alkaline or heavy duty

● Secondary cells or rechargeable batteries;

   Lead-acid battery
    -vented lead-acid
    -automotive (car)
    -deep-discharge or traction
    -stationary
    -low-antimony solar battery
    -sealed or valve-regulated

   Nickel-Cadmium batteries
    -vented
    -sealed.

Figure 7.1   Battery-charging workshop in downtown Nairobi. Most of these batteries are used for powering lighting systems and TVs in surrounding semi-urban areas which are not connected to the grid

7.2 Primary Cells (Dry Batteries)

The familiar torch (flashlight) battery is perhaps the most common power source for small portable electric lamps. This battery type comes in various standard sizes as illustrated in Figure 7.2. Although the purchase or first cost of dry cells is relatively low, paradoxically, it is one of the least cost-effective electrical power sources in terms of the cost per unit of useful energy delivered. Furthermore, only a limited energy yield can be obtained before the battery has to be thrown away.

Dry batteries or primary cells are used in large numbers especially by the rural poor, mainly because they are convenient, just about affordable and generally all that is available. However their high cost makes them only suitable for powering small portable lamps that can only be used economically for short periods; e.g. torches for finding the way in the dark, for hunting or for emergency use when another form of lighting fails.

Primary cells are based on an irreversible electrochemical reaction and consequently cannot be recharged. Once the chemicals are exhausted the battery is useless and must be disposed of.

In some cases there are distinct environmental concerns about the disposal of used cells

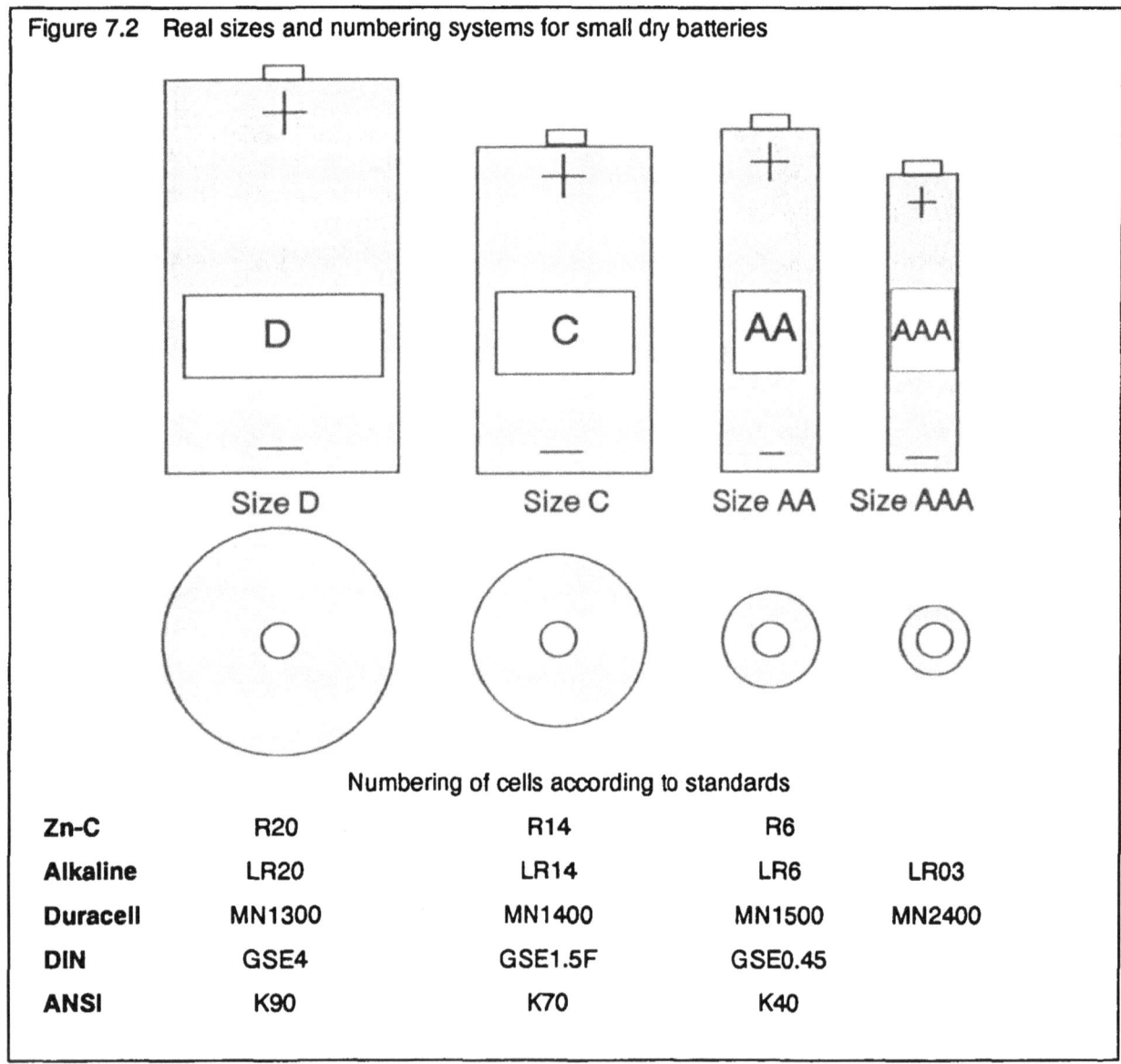

Figure 7.2   Real sizes and numbering systems for small dry batteries

Size D   Size C   Size AA   Size AAA

Numbering of cells according to standards

Zn-C	R20	R14	R6	
Alkaline	LR20	LR14	LR6	LR03
Duracell	MN1300	MN1400	MN1500	MN2400
DIN	GSE4	GSE1.5F	GSE0.45	
ANSI	K90	K70	K40	

since some of the chemicals present in dry cells are potentially harmful (mercury in particular). Battery manufacturers are presently trying to find less objectionable alternative chemicals to make their products less hazardous when disposed of ('green' batteries). Nowadays, the mercury content of most dry cells has been significantly reduced, if not eliminated by most of the more reputable manufacturers in developed countries. However, battery recycling is still necessary in order to eliminate the concerns about chemical toxicity and depletion of metal resources.

In recent years, primary cell technology has improved and usually two distinct qualities of cell are available in any size: standard Zinc carbon (i.e. the basic type) and Alkaline (i.e. the 'heavy duty' or 'long life' type).

Zinc-carbon cell

The most widely used and cheapest form of primary cell is the Leclanché cell or zinc-carbon cell as illustrated in Figure 7.4. An outer zinc container forms the negative pole

---

Figure 7.3 Cell, battery and connection definitions

CELL

The basic unit that stores electricity is called a cell. A single cell produces either 1.5V (if zinc-carbon), 1.2V (if nickel-cadmium) or 2.0V (if lead-acid). Since many lighting applications need anything from 3 to 12V or even 24V, it is normal to connect cells in series so their voltages add up to the required value. For example, 6V can be achieved by connecting in series three 2.0V lead-acid cells or five 1.2V nickel-cadmium cells.

BATTERY

A packaged combination of cells is technically known as a 'battery'. In most cases a number of cells are packaged in a single container or sleeve, typically three or six 2V lead-acid cells to give a 6 or 12V battery.

SERIES CONNECTION

If cells or batteries are connected + to - (i.e. positive of one cell to negative of the following cell) so that their voltages add to a suitable value for the application, they are said to be 'series connected'. All cells will have the same current passing through them.

PARALLEL CONNECTION

If two or more cells or batteries are connected + to + and - to - (i.e. positive of one cell to the positive of another and similarly for the negative poles) then they are said to be 'parallel connected'. Two cells connected in parallel would produce the same voltage as a single cell, but be capable of delivering twice the current. They would also have double the electrical storage capacity of a single cell at the same voltage.

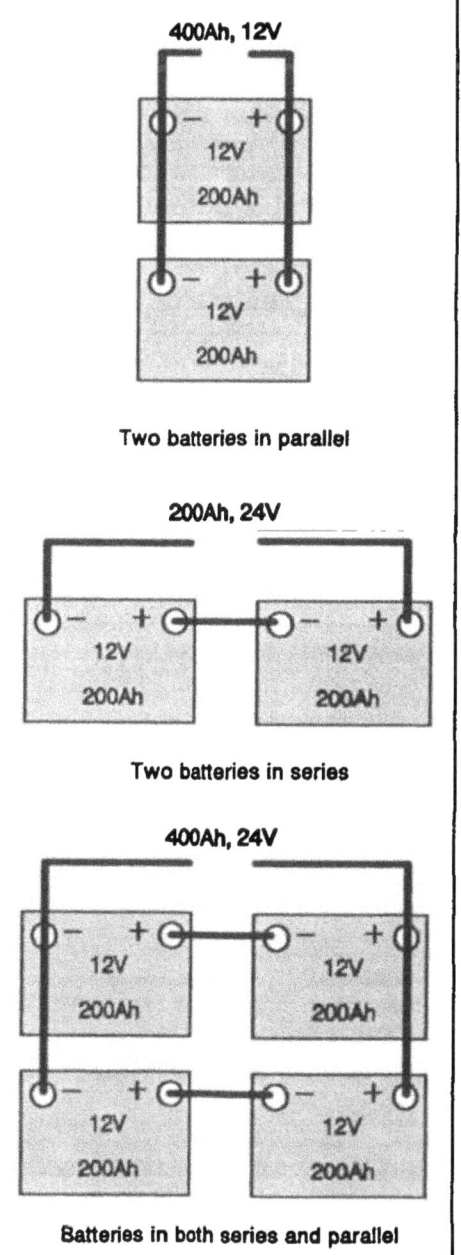

Two batteries in parallel

Two batteries in series

Batteries in both series and parallel

---

and a central carbon rod, normally fitted with a metal button on the outside, forms the positive terminal.

Operating principle

Zinc-carbon cells derive their energy from a chemical reaction which takes place whenever they are connected to a closed electrical circuit. A layer of ammonium chloride paste provides an electrolyte, and manganese dioxide (with graphite added to improve conductivity) surrounds the carbon rod to absorb hydrogen that would otherwise be given off and which would reduce the efficiency of the cell. When a conducting electrical circuit is made between the poles of the cell, chloride ions from the ammonium chloride paste give up electrons to the zinc. This gradually dissolves the zinc by forming zinc chloride. The electrons released then flow around the circuit as electricity and produce a mixture of ammonia and hydrogen when they reach the interface between the carbon rod and the electrolyte paste. Eventually the zinc corrodes away to such an extent that the cell ceases to work.

The voltage of any zinc-carbon cell is always 1.3 to 1.5 volts when the chemicals are fresh. The size of the cell only influences the current (and hence the power) that can be produced.

Alkaline cells

Alkaline cells are in fact more sophisticated in design than a basic zinc-carbon Leclanché cell and have a much larger electrical capacity (see Table 7.1). It is beyond the scope of this book to go into detail on the specific electro-chemical processes, which in principle are not dissimilar from the basic zinc-carbon cell. Alkaline cells are also called manganese dioxide cells or 'heavy duty' batteries. Their open voltage is 1.5 volts when the chemicals are fresh.

Electrical characteristics

As a cell discharges its voltage falls. For example, a fresh zinc-carbon cell may have an open voltage of 1.5V, but towards the end of its life the voltage will fall to around 0.8 to 0.9V (see Figure 7.5). At this point, the useful

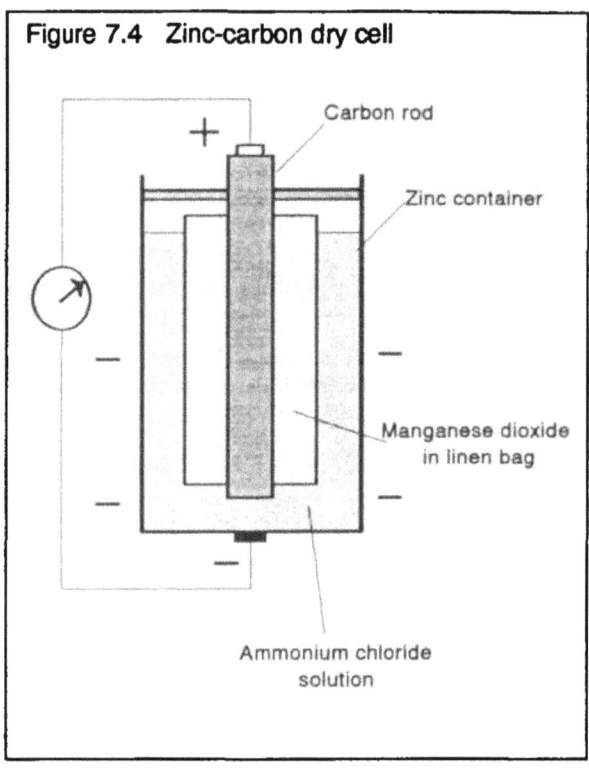

Figure 7.4   Zinc-carbon dry cell

Carbon rod

Zinc container

Manganese dioxide
in linen bag

Ammonium chloride
solution

discharge is considered to be complete (e.g. a standard bulb won't give any more light if connected to the cell). The same rule applies to all forms of batteries and cells; the voltage falls with the state of charge of the cell.

The electrical capacity is the total quantity of electricity (i.e. quantity of energy) that a cell can deliver. The electrical capacity depends on many factors, such as cell size, cell type, rate of discharge, temperature and mode of use. For a given type, the bigger the cell, the higher the electrical capacity (see Table 7.1).

Influence of the rate of discharge

Figure 7.5 indicates typical C-sized primary cell characteristics at two rates of discharge. One is about 0.5A (i.e. 1.1V/2.2ohm = 0.5A). This corresponds approximately to an average of about 0.55W (i.e. 1.1V x 0.5A) per cell. The other of only 0.3A (1.1V/3.9ohm = 0.3A) corresponds approximately to a mean of about 0.33W (i.e. 0.3 x 1.1V). It should be noted that the life of the cell is more than three times as long at the lower rate of discharge.

Also, as the output averages approximately 0.55W (in reality it will start at a higher level and gradually reduce to a lower level) then the

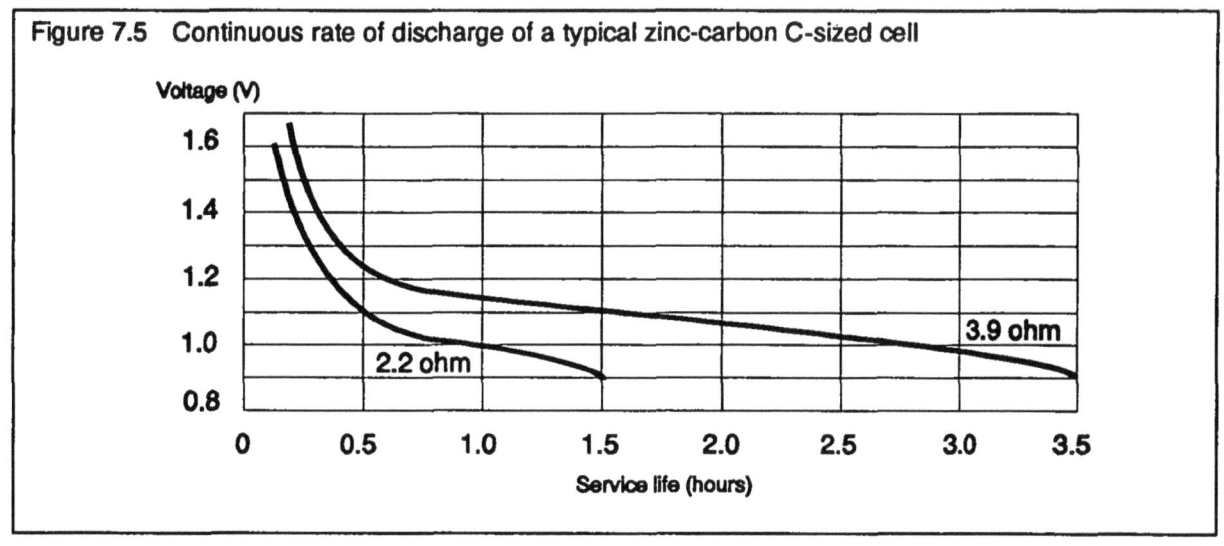

Figure 7.5   Continuous rate of discharge of a typical zinc-carbon C-sized cell

energy delivered will be approximately 0.55W x 1.5h = 0.82Wh, whereas at the lower rate of average discharge, it will be approximately 0.33W x 3.5h = 1.15Wh, which is nearly 40% more energy from the same type of cell.

In other words, the cell has a higher capacity at the lower rate of discharge. Therefore the true capacity of a cell is not a fixed quantity but is much greater at lower levels of discharge current. Hence, with primary cells, it is recommended that, whenever possible, as small a current as possible is used.

The power requirements of torches or flashlights make quite substantial demands on cells. Thus typical flashlights will draw 0.3 to 0.5 amperes from a D cell and 0.2 to 0.3 amperes from C and AA cells. Under these conditions, less than the rated capacity of the cells is obtained. In typical intermittent service, the decrease in capacity can be 50% for a D size cell, and even more for the smaller cells. On continuous drain, the situation is even worse; the D cell may deliver only 25% of their rated capacity.

In order to optimize the use of dry cells, it is a common practice among the poorer rural communities in developing countries to use dry cells for radios and cassette players until their voltage falls to such a low level that they do not function any more (most electronic devices need a certain minimum voltage to function at all), and then the partially used cells are 'finished off' in torches where low voltage simply results in a rather dim and yellow light.

Influence of temperature

The capacity of dry cells, like most other batteries, increases with higher temperatures. Usually the capacity is given at 20°C. Above this temperature the capacity is increased and vice versa. This explains why in some cases, the user may warm the batteries to get extra power from them (e.g. by exposing the batteries to sunshine).

Self-discharge and shelf life

Primary cells have the advantage of being stable in terms of self-discharge. In other words, it is possible to keep them on open circuit (i.e. switched off) for a long period without serious loss of capacity. The 'shelf-life' of primary cells varies, but some of the better quality alkaline 'heavy duty' types can be kept several years with no more than a few percent loss of capacity. Some of the better quality ones marketed today have their 'use by' date printed on them, and are sold in factory-sealed packs, which is a good precaution against being sold old and possibly degraded units. The cheaper lower-powered kind (i.e. standard zinc-carbon type) tend to deteriorate more quickly, but even so they retain their capacity better than any other type of portable electrical power source.

The self-discharge rate is adversely affected by storage at high temperatures, so dry-cells should ideally be stored at temperatures in the range 10-25°C and at a relative humidity of below 65%. Unfortunately, the most common types of primary cells manufactured in developing countries tend to be of the lowest quality by international standards and are often stored in unfavourable conditions. Hence their capacity can often be below standard, giving even worse value for money.

Electricity cost

The cost of electricity from primary cells typically works out to something in the range of US$140 to $1300 per kWh (which is about 700 to 6500 times more expensive than mains electricity taken at US$0.2 per kWh). Therefore although the initial cost of a primary cell is small, the unit cost of electricity from it is extremely high.

It generally does not take long for a torch user in a remote location to spend more on batteries than the purchase cost of the torch. But, because the expenditure is usually spread over a period of time, many users are not really aware of how expensive a form of lighting this is. It is also unlikely that they have an appropriate alternative to serve their needs. This is another example of the poor having to use one of the most expensive solutions to their problems, due to lack of capital.

Table 7.1 indicates the typical costs and capacities of the main types of standard primary cell. The costs given in the table are merely indicative.

Table 7.1	Typical specifications for dry batteries				
Size	Open voltage (V)	Rated capacity (Ah)	Useful[1] energy (Wh)	Unit price (US$)	Energy cost[2] (US$/kWh)
ZINC-CARBON (basic type)					
AA / R6	1.5	0.9	0.3[3]	0.40	1300
C / R14	1.5	1.8	0.7[4]	0.60	850
D / R20	1.5	5.5	2.2[5]	1.00	450
ALKALINE MANGANESE TYPE (long-life/heavy duty)					
AA / LR6	1.5	2.2	1.9[3]	1.00	520
C / LR14	1.5	7.2	6.5[4]	1.70	260
D / LR20	1.5	16.0	14[5]	2.00	140

NOTES

1   Useful energy (Wh) has been obtained from typical characteristic curves of dry cells and for a load typical for lighting applications (e.g. torches)

2   The cost in US$/kWh is calculated as the price of cells divided by the capacity in kWh. The cost is only valid for a given rate of discharge corresponding to the above application.

3   discharge rate 0.25A (2 cells in series for a 0.6W/2.5V bulb)

4   discharge rate 0.25A (2 cells in series for a 0.6W/2.5V bulb)

5   discharge rate 0.5A (2 cells in series for a 1.2W/2.5V bulb)

IMPORTANT: The price of dry cells in developing countries varies greatly depending on whether or not they are locally manufactured. For example, a D size dry cell costs only US$0.30 (1991) in Zaire. Hence the costs given in this table are merely indicative.

7.3 Secondary Cells (Rechargeable Cells and Batteries)

There are for most practical purposes two main types of secondary cell, namely lead-acid and nickel-cadmium (NiCd). Other types exist. But they are either for specialized purposes or still under development and so are not at present readily available and in general use.

Before going into more details, there follows a summary of terms used to characterize secondary cells of which the reader should be aware. Manufacturers publish details of their products using these terms, so an understanding of their meaning is quite useful when referring to their literature.

Terms used to specify rechargeable cells and batteries

Rated capacity (C) [unit Ah]

A battery's rated capacity (C) is the total quantity of electrical charge (i.e. current x time) in Ah (ampere-hours) that can be drawn from a fully charged state at a specified discharge rate and electrolyte temperature before the voltage falls to a specified cut-off voltage.

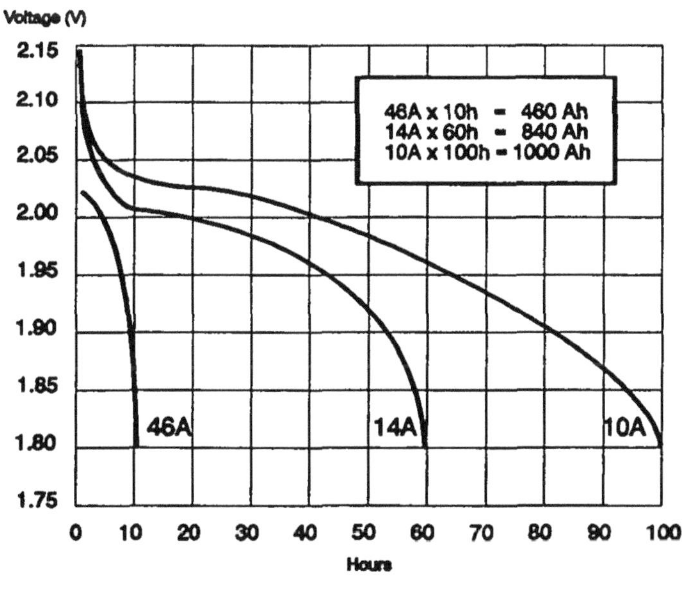

The rated capacity varies with the temperature and varies strongly with the discharge rate. The higher the discharge rate, the lower the capacity (see Figure 7.6). The rated capacity should always be specified using the combined factors: discharge rate, temperature and cut-off voltage (e.g. a lead-acid battery of 100Ah at C/10, 25°C and 10.8V cut-off voltage).

Example: a 200Ah (at C/10) battery is able to provide 20A current for 10hrs before it is completely drained of energy (i.e. 200Ah/10 x 10h = 200Ah).

Discharge rate (C/n) [unit A]

This is the discharge current which is expressed as a fraction of the rated capacity. For example a discharge rate of C/10 is 12A for a 120Ah rated battery.

State of charge (SOC) [unit %]

The state of charge is the amount of charge left in the battery expressed as a percentage of the rated capacity (i.e. 100% is full charge, 50% is half charge).

Usable capacity (Cu) [unit Ah]

This is the fraction of the rated capacity that is used for each cycle (usually shallow cycles) to allow a reasonable cycle life. The usable capacity should always be greater than the average need for electricity to prevent over-discharging.

Depth of discharge (DOD) [unit %]

This is the ratio, expressed as a percentage, of the usable capacity to the rated capacity of the battery. So a low DOD implies that less charge is taken from the battery at each cycle. For instance, if the rated capacity is 100Ah and the DOD is 30% then the usable capacity is 30Ah. The lower the DOD that is selected, the longer the battery life will be. The allowable DOD is influenced by temperature so

Figure 7.6 Typical variation of battery capacity with rate of discharge

46A x 10h = 460 Ah
14A x 60h = 840 Ah
10A x 100h = 1000 Ah

that in low ambient temperatures the DOD should be reduced.

When the DOD is greater than 50%, the battery is 'deep cycling'. When the DOD is smaller than 20 to 30%, the battery has a 'shallow' cycle. Shallow cycling of batteries is always preferable.

The rated capacity is a characteristic of the battery, the usable capacity is defined by the daily demand of the load and the DOD is selected by the designer. As an example, if the electricity need is one 13W light for 10 hours per night, i.e. 130Wh or 11Ah with a 12V battery, the usable capacity should be at least 11Ah. If the DOD is 20% to allow, say, a life of 1000 cycles, the minimum rated capacity needed is 55Ah (i.e. 11Ah/20%).

Charge rate (C/n) [unit A]

This is the current at which a battery should ideally be charged. Charging a secondary cell or battery at too high a rate can cause permanent damage to the battery. As a rule of thumb, most batteries (including nickel-cadmium) can be safely charged with a current equal to one tenth of the rated capacity (i.e. C/10). For example, a 50Ah battery can be safely charged at 5A.

Self-discharge [unit % / month]

The self-discharge is the charge lost expressed as a percentage of the initial state of charge when the battery is not used over a period of one month.

If left unused and uncharged, all batteries slowly lose their charge, but some types self-discharge faster than others (see Table 7.2). Be aware that internal self-discharge increases with increased temperature.

Charge/discharge efficiency [unit %]

Charge/discharge efficiency is the ratio (expressed as a percentage) of the amount of energy used by the loads (e.g. lights) to the amount of energy needed to recharge the battery fully over a complete cycle. A low efficiency means that more charging energy will be needed for a given daily electricity demand.

Cycle life

This is the number of times (i.e. cycles) a battery can be charged and discharged before it permanently loses more than 20% of its rated capacity. The cycle life varies with the discharge rate, depth of discharge and temperature.

For example, with a photovoltaic charging system the battery will be recharged every day when the sun shines and discharged every night to run lights, so the cycle will be 24 hours. If after one year (365 days or cycles) the battery capacity were to have fallen to just less than 80% of its rated capacity, its cycle

Figure 7.7
Battery-recharging workshop in Nairobi city centre. Note that the technician is wearing plastic gloves to protect himself from acid attacks. These batteries may have travelled several kilometres on a truck or bicycle before being charged to power the lights, radio or TV of the happy owner.

life would be 365. If a battery were taken to the nearest town to be recharged once a week from the mains, then it would be on a weekly cycle, and if the battery lasted one year with more than 80% of its rated capacity, its cycle life would be 52.

The cycle life is always specified using the combined factors of: depth of discharge (DOD), rate of discharge (C/n, n is usually equal to 10 or 100) and temperature (e.g. 1000 cycles at 80% DOD, C/10 and 25°C).

When a battery is rarely cycled but always kept charged (e.g. emergency lighting), the capacity is inherently reduced with the time. In this case a 'calendar life' is defined and measured in years.

Lead-acid batteries

The least expensive option for any significant size of electrical battery storage is the lead-acid battery.

Operating principle

A lead-acid battery consists of lead and lead oxide electrodes that interact with sulphuric acid in an electrochemical process which is reversible. This in turn results in a flow of electrons through the battery and into the load. The electrolyte is a solution of sulphuric acid in water (a highly corrosive liquid which

is poisonous and attacks the skin, destroys clothes and many other things). The electrolyte gets more diluted as the state of charge is reduced. Hence, a fully discharged battery has a weak solution of electrolyte and in sub-zero ambient temperatures is in danger of freezing and being damaged.

Vented (unsealed) and sealed lead-acid batteries

Traditional designs of lead-acid battery (i.e. vented or un-sealed) need occasional replenishment with distilled water to maintain the correct strength of dilution of the acid as some water will be lost with normal use. However, modern designs of lead-acid battery are sealed and have the acid electrolyte in a 'dry' form as in a primary dry cell.

Plate construction

In a lead-acid battery, the powdered lead (negative plate) and lead oxide pastes (positive plate) are dried onto an electrode grid to form a plate that holds the material in place while acting as collection grid for the electrons that flow in the battery. As pure lead and lead oxide are not strong enough, antimony, calcium and other substances are added to create lead-alloys which are stronger and therefore increase the life of the battery.

The presence of antimony increases the cycle life with deep-cycling, but on the other hand

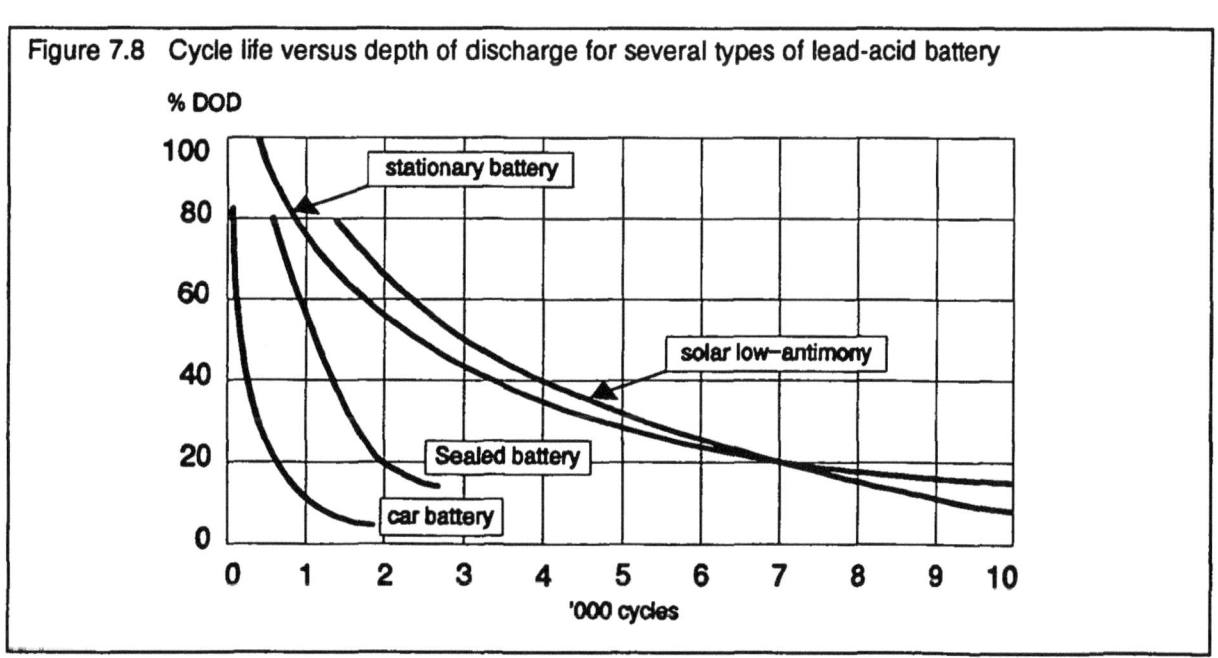

Figure 7.8 Cycle life versus depth of discharge for several types of lead-acid battery

increases the distilled water consumption and self-discharge. Calcium generally allows a lower distilled water consumption allowing infrequent or no maintenance, but greatly reduces tolerance to deep-discharge.

Finally, the thicker the positive plates, the better the cycle-life when deep cycling. The positive plate thickness may vary from 1.5mm to more than 6mm depending on the type of battery. The choice of an appropriate thickness and chemical composition of the plates defines the performance of a battery. Therefore, before purchasing a battery its main constituents should be known.

General characteristics

Lead-acid batteries have a nominal fully charged voltage of 2V per cell. So a 12V battery typically has six cells in series. The charge-discharge efficiency for a lead-acid battery is typically 70%. The energy density is about 35Wh/kg of battery.

A lead-acid battery will only withstand a certain number of charge-discharge cycles before it fails and needs to be replaced. The greater the depth of discharge (i.e. the more on average that the battery is 'flattened') the fewer cycles it will survive, and vice versa. Figure 7.8 illustrates this and shows how, in this example, if the battery (e.g. a stationary lead-acid battery) is discharged regularly by 80% of its total capacity it may last 800 cycles but if discharged only 20% it may last 6000 cycles. If the battery were discharged at 20% rather than 80%, the rated capacity will have to be four times larger to deliver the same energy, but will last at least four times as long.

Therefore the size of a battery is a compromise between making it so large that it is too expensive to be affordable or so small that it gets discharged too much by the average demand for electricity, and hence has too short a useful life.

Battery capacities are usually specified for 25°C operating temperature. The capacity is typically reduced by 1% per 1°C going down to 0°C, but increases approximately 1% per 1°C going from 25°C to 40°C. On the other hand, the life of a battery (taken at 25°C) decreases strongly with the increase of temperature. Hence in a tropical climate, a battery should be kept whenever possible in a cool and well ventilated room.

Lead-acid batteries can be sub-divided into several categories, but to simplify, five categories are considered here, the first four being vented types:

- Automotive (vehicle);

- Deep-discharge or traction;

- Stationary;

- Low-antimony solar battery;

- Sealed or valve-regulated battery.

Automotive batteries

These batteries have been designed for starting cars and lorries. Automotive batteries have a poor capacity for their size and a poor cycle life. The plate alloy composition varies greatly. Some have a high content of calcium (for maintenance-free, or almost free, car batteries) and some have a few percent of antimony. A typical automotive battery will

Figure 7.9 Checking the state of charge of a battery with a hydrometer at a solar charging station in Nselo, Zaire

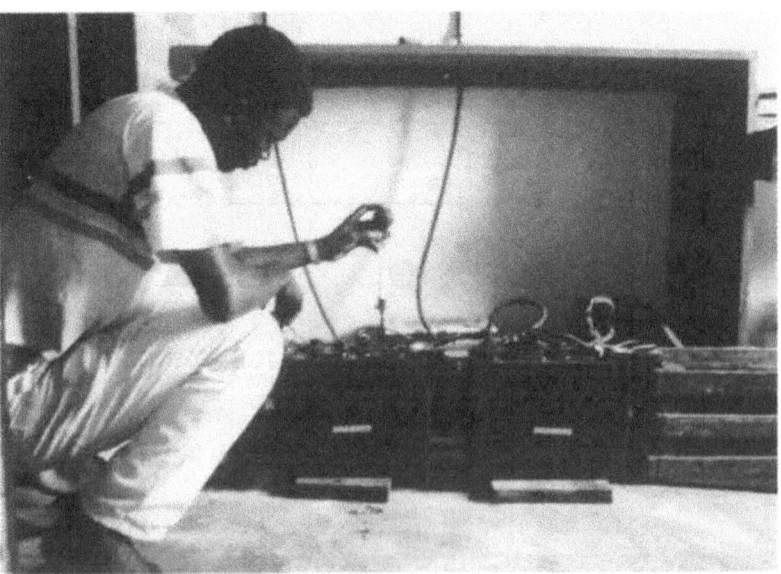

only withstand about 20 deep-discharge cycles before it becomes completely useless. Car batteries are easily damaged if left discharged for any length of time, since, normally when used in a car, every time they are discharged they are immediately recharged. The cell design in a car battery is optimized to deliver heavy currents (having large areas of thin plates) and therefore it is poorly suited to supplying smaller currents for many hours before being recharged, as is required for most lighting applications.

Figure 7.10 Low-antimony solar battery bank, Likasi, Shaba in Zaire

In spite of the short life when used for lighting applications, automotive batteries have the advantages that they are usually the cheapest batteries when compared by rated capacity (typically between 50 to 100Ah), they are often locally produced and are widely available and reparable (i.e. one or several damaged cells can be replaced in local workshops or at the local factory). This is why automotive batteries are used in developing countries for deep-cycle applications where better batteries are not available and/or are too expensive. Ideally if a car battery is used for lighting, it should not be discharged more than about 30% of its rated capacity nor should it be left in a discharged state for long before it is recharged.

Automotive batteries for trucks and buses are slightly better than the car batteries because their thicker plates are designed to withstand more vibrations and deeper cycling. However they are also heavier and more expensive but may be appropriate for large lighting systems.

Deep-discharge or 'traction' batteries

Deep-discharge batteries have a high antimony content (e.g. 4 to 8%) and can tolerate discharge to as much as 80% of their rated capacity with a cycle life from 1000 to 1500 deep cycles. However, they tend to lose water at a faster rate than other types and need frequent maintenance. They are commonly used for electric vehicles and are often known as traction batteries. The self-discharge rate is also fairly high. Deep-discharge storage

batteries are available with flat or with thick tubular positive plates (the latter tend to be for heavier duties and are more expensive). Access to each cell is provided to give the electrolyte an occasional top-up with distilled water.

Deep-discharge batteries may be suited for lighting applications on technical considerations but are relatively expensive, require a lot of maintenance and are not often locally available.

Stationary batteries

These batteries are often called stand-alone or standby batteries and have been designed to supply power when there is a grid failure. In most applications (e.g. emergency systems for telecommunications, etc.) they are kept fully charged by the mains supply and are ready to take the load whenever needed. They are extremely reliable, have a low self-discharge rate and a long cycle life with shallow cycle (i.e. they can last over ten years). Stationary batteries have thick low-antimony positive plates, a large reservoir of electrolyte and are extremely reliable. When used for stand-alone applications, the stationary battery bank is oversized so that it only runs with shallow cycles. Second-hand stationary batteries may be well suited for lighting applications.

Low-antimony solar batteries

These batteries, which are quite similar to stationary batteries, have been specifically

79

designed for photovoltaic systems and are therefore appropriate for electrical lighting systems. Low antimony and thick positive plates give a high cycle life for deep cycles. The low percentage of antimony means that the distilled water consumption is low and the self-discharge rate is below 3%. The loss of water can be reduced to a very low level by direct recombination caps mounted on top of each cell. Usually the casing is transparent which allows visual inspection of the level of the electrolyte. In most cases, there is a large reservoir which reduces the frequency of topping-up still further. The cycle life of solar batteries ranges typically from 1200 at 80% DOD to 3000 at 50% DOD. They are usually delivered dry-charged and the electrolyte is added at the site when the battery is being installed. These batteries are fairly expensive and available only from photovoltaic systems suppliers.

Sealed or valve-regulated batteries

Most sealed or valve-regulated batteries use lead alloyed with calcium instead of antimony in the positive plates in order to reduce the gassing to a minimum. The electrolyte is either absorbed onto a 'glass mat' or is in a gelled form so that it cannot be spilled and contains chemicals that absorb the hydrogen and oxygen that is produced. The recombination process is not 100% efficient and so some gasses are emitted, but only 1% of that of an unsealed battery.

The gasses are evacuated through a regulated safety valve, but the battery contains enough electrolyte for its entire life and therefore does not need any maintenance. Hence they are often refered to as 'maintenance-free' batteries.

Because of their calcium content, sealed batteries have a short cycle life for deep cycles. They have a low rate of self-discharge and can support a full discharge, but must be recharged as soon as possible to prevent permanent damage.

Overall, a sealed battery is likely to have a shorter life than a well-maintained unsealed battery with the same alloy contents, but will obviously last longer than a poorly maintained unsealed battery.

Figure 7.11 Small sealed lead-acid battery

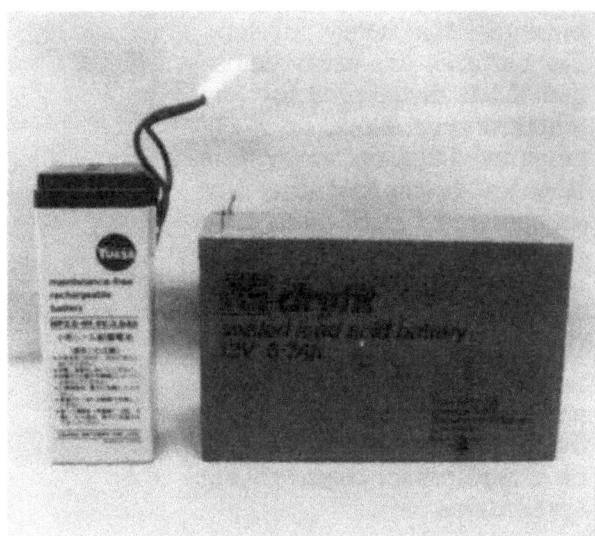

The main disadvantage of sealed lead-acid batteries is their need for regular recharging to prevent sulphate build-up leading to permanent damage during transport and storage prior to installation. Batteries would typically be recharged in storage every three months. This can present further problems in developing countries where the high ambient temperatures accelerate the self-discharge.

Sealed batteries have recently been developed that contain a mixture of calcium with antimony with a gelled electrolyte, and have a good cycle life with deep-cycle use (e.g. up to 800 cycles at 80% DOD). This makes them very competitive in comparison with vented low-antimony solar batteries. They are likely to be well suited for lighting systems (particularly portable systems) as they do not require maintenance.

Charging of lead-acid batteries

Lead-acid batteries ideally need to be recharged in a carefully controlled manner. This is accomplished by a specially designed DC power source that contains sensing circuits (for voltage and temperature of battery) to adjust or stop the charge rate automatically as appropriate. The final charging voltage decreases linearly with the battery temperature (typically 15.2V at 0°C and -0.03V/°C above 0°C, e.g. 14.3V at 30°C). Normally a high rate of charge is acceptable when the state of charge of the battery is low, but to obtain best results and a long life from the battery, the charging current needs to be

Table 7.2   Principle characteristics of various batteries

Type	Depth of discharge	Self-discharge (capacity per month)	Cycle life	Calendar life (cell life)	Approx. cost (<100Ah)	Approx. cost (>100Ah)
	%	%	no. cycles	years	US$/kWh	US$/kWh
LEAD-ACID						
Automotive	20	30	300-600	1-3	100-150	80
	80		20			
Traction	80	5-7	1500	4-6	200-400	200
Stationary	50	3	3000	5-10	300-400	250
	80		1200			
Solar	50	1-3	3000	5-10	250-350	200
low antimony	80	3	1200			
Sealed	20	2-6	400-1500	4-8	150-500	200
	50	2-6	400-1000	4-8		
NI-CAD						
Sealed	100	5-30	100-1000	3-5	600-1000	N/A
Unsealed	100	3-5	1000-2000	20	5000	350

NOTE   the cost in US$/kWh is calculated as follows: price of battery divided by rated capacity.

tapered off as the battery approaches its fully charged condition.

Batteries can often benefit from an extended charge at low current levels (called a trickle or equalization charge) to equalize the voltages of all cells. This is because sometimes one cell may not accept its charge at the same rate as the others and so it is left only partially charged while the others are fully charged. In this case the partially charged cell may become over-discharged through use, and eventually fail prematurely compared with the other cells.

Overcharging or charging at too high a rate (i.e. too great a current) causes hydrogen bubbles to be generated on the battery plates which can cause damage to the battery's internal structure. Therefore it is important to adhere to manufacturers' recommendations on battery charging. With a standard 12V lead-acid battery, it is usual to charge it with a current of C/10 at around 14V at 25°C. It can be considered discharged if the open circuit voltage falls below about 11.5V,

assuming it has not delivered a heavy current in the immediate past.

Measurement of the state of charge

There are two ways to check the state of charge of a lead-acid battery. The most reliable, if the electrolyte is accessible, is to use a hydrometer (available from motor vehicle accessory suppliers) to suck up a sample of the battery acid and measure its specific gravity. (Do not forget to empty the electrolyte back into the same cell). Figure 7.12 illustrates a typical correlation between electrolyte specific gravity, cell voltage and state of charge. Note that the specific gravity of the electrolyte depends on the make and type of battery and also on the temperature of the electrolyte. In order to make an accurate measurement, it is necessary to correct the reading with appropriate tables usually given by the battery manufacturer or to use correction factors. The manufacturer's data (i.e. state of charge vs specific gravity) is usually given for a battery temperature of 15°C. Therefore, in the absence of conversion tables a correction factor of 0.007 should be added to the hydrometer reading for every 10°C above 15°C and subtracted for every 10°C below 15°C. The resulting figure can then be used with the manufacturer data. For example if the temperature is 25°C and the hydrometer reading is 1.220 then equivalent specific gravity at the 15°C reference temperature would be 1.220 + 0.007 = 1.227.

Batteries in tropical climates should be filled with electrolyte of a lower specific gravity than in temperate climates (e.g. 1.240 when fully charged and 1.110 when fully discharged at 25°C).

The method of checking state of charge used by most automatic battery charge regulators is to sense the battery voltage. Unfortunately this is not a completely reliable measure of the battery state of charge as the terminal voltage is affected not only by the state of charge, but by the temperature of the battery and its immediate 'charge-discharge history'. The charge-discharge history can be illustrated as follows: if a heavy current has been drawn, even briefly, it can depress the battery voltage for some time giving an apparently lower state of charge than really exists; conversely, if a heavy charge current has been applied, it can raise the voltage above that which would apply under steady-state conditions.

Consequently, the only reliable way to check a battery by its voltage is to measure the open-circuit voltage after the battery has been standing idle (i.e. with no electrical load) for some time, preferably more than 30 minutes.

Both the open voltage of the battery and the specific gravity of the electrolyte should be recorded in order to create a history file for the battery. This is one of the aspects of a good maintenance programme.

The measurement of the voltage and specific gravity are not always enough to assess whether or not a battery is healthy or needs to be repla-

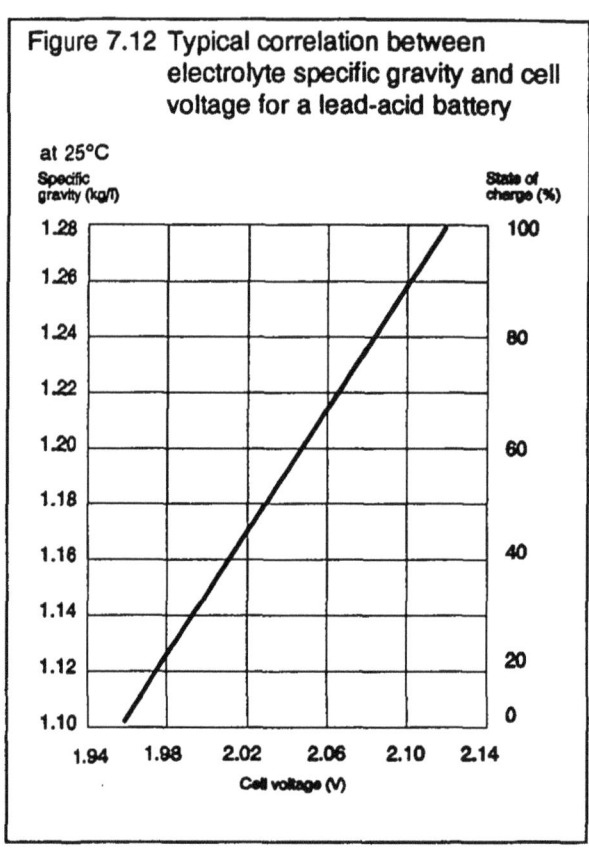

Figure 7.12 Typical correlation between electrolyte specific gravity and cell voltage for a lead-acid battery

ced. A load tester is then necessary, which applies a very high load (typically a few hundred Amperes) for a few seconds and measures the voltage drop. If this apparatus is not available, it is advisable to carry out a test where the battery is fully charged and then discharged through a load (e.g. at C/100) until the voltage cut-off is reached (i.e. around 11.5V at 25°C).

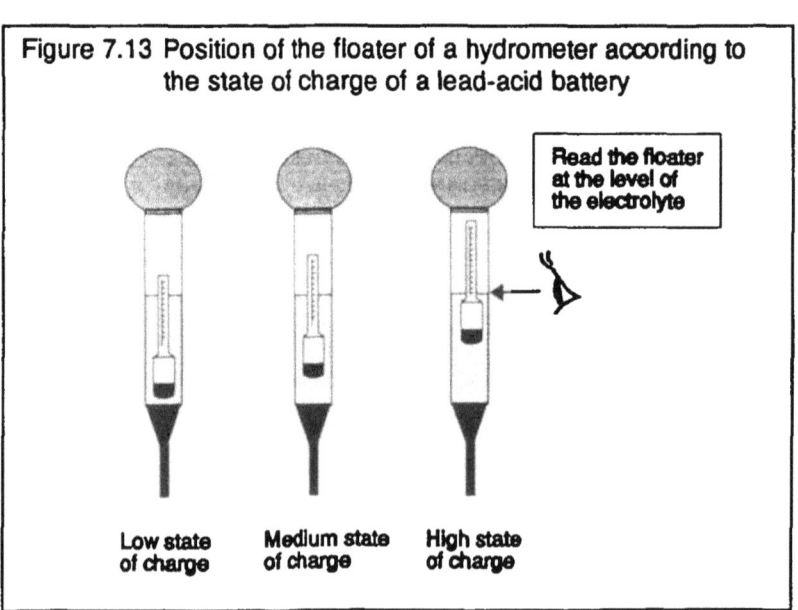

Figure 7.13 Position of the floater of a hydrometer according to the state of charge of a lead-acid battery

SAFETY AND ENVIRONMENT HAZARDS OF LEAD-ACID BATTERIES

Vented batteries: Care is obviously needed as, apart from the battery acid being extremely corrosive, hydrogen gas is produced which is highly flammable and potentially explosive when mixed with air. Thus care should also be taken to avoid naked flames or sparks in the battery enclosure, especially if the battery is housed in a confined space. NEVER check the electrolyte levels with a naked flame such as a kerosene lamp or a candle. For the same reason, battery storage areas should be well ventilated.

Sealed batteries: These contain the electrolyte in 'dry' form so that no electrolyte can be spilled, and so there is less of a hazard. Even so, care must be taken not to damage the casing.

Recycling: Both types of batteries should be disposed of safely. Where practical, it is a good idea to give away lead-acid batteries to local battery manufacturers for lead and plastic casing recycling.

Nickel-cadmium batteries

The main alternative to the lead-acid battery is the nickel-cadmium or 'ni-cad' battery. Like lead-acid, ni-cad batteries are available either vented (with sintered or pocket plates) or sealed. Vented ni-cad are designed for applications which require robust energy storage with long operating lifetimes and minimal maintenance. Sealed and usually small (i.e. sized AAA, AA, C or D) ni-cad are used as an economical replacement for dry cells.

Operating principle

Ni-cad cells use a nickel hydroxide anode and a cadmium cathode with potassium hydroxide (a strong alkali) as the electrolyte. As the cell is discharged the anode is converted to nickel hydroxide and the cathode to cadmium hydroxide. The electrolyte acts only as a medium to transport ions and its specific gravity remains unchanged regardless of state of charge. Thus specific gravity cannot be used as a state of charge indicator.

Main characteristics

The nominal voltage of a ni-cad cell is 1.2 volts, so a nominal 12V ni-cad system needs 10 cells. Typical specific energies of ni-cads fall in the range of 20 to 45Wh/kg. ni-cad cells can withstand a greater depth of discharge than a lead-acid battery, and so generally a smaller capacity can serve a given duty. They also tend to last longer (e.g. 10 to 20 years for the larger ones). Also, nickel-cadmium batteries are less easily damaged by over-discharge or over-charging and so simpler and cheaper charge control systems can be used to compensate for their extra unit costs. In addition they are more tolerant of extreme temperature variation than lead-acid batteries and can therefore operate at sub-zero temperatures.

Although ni-cad batteries are robust and reliable, they do have a few shortcomings that can cause problems. One major problem is that reversing the polarity when recharging a ni-cad cell usually destroys it completely. This can happen sometimes not because a cell was reversed by carelessness when wiring it up for recharging, but when one

Table 7.3 Capacity of small ni-cad cells versus size				
Size	AAA	AA	C	D
Voltage (V)	1.2	1.2	1.2	1.2
Capacity* (Ah)	0.18	0.5-0.75	1.2-2.2	2.2-4.4

* The capacity is given at C/10.

NOTE: When buying ni-cad cells, it is important to specify both the size and capacity as for instance a size D battery may have a capacity varying from 2.2 to 4Ah depending on its construction.

cell in a battery of ni-cad cells is weaker than the rest: then the good cells can cause reverse charging of a weak one in certain circumstances, thereby destroying the weak one completely. This is one reason why it is not a good policy to mix old cells and new ones either for recharging or for actual use.

Another characteristic of ni-cad batteries is a tendency to self-discharge rather more quickly than lead-acid cells and much more quickly than primary cells. Therefore ni-cad primary cell substitutes need regular recharging and are less useful for occasionally used loads than for regularly used ones. They are therefore well suited for small photovoltaic applications where they are being charged with the daily sunshine.

> CAUTION: Ni-cad batteries should be disposed of carefully to avoid cadmium pollution.

Memory effect of ni-cad batteries

The memory effect is the tendency of a battery to adjust its electrical properties to a certain duty cycle to which it has been subjected for an extended period of time. Vented pocket plate batteries do not develop the memory effect.

Figure 7.14 Small sealed ni-cad batteries (left: AA sizes, right: C sizes)

However, sealed sintered plate cells, for example the AAA, AA, C and D sizes have a memory effect. To remedy to this problem, as they have been used in a given cycle or 'asleep' on the self, they need to be 'awakened' by being fully charged and discharged at C/10 for 3 or 4 cycles before their memory is 'stretched' enough to hold a full charge. If only light charging is applied (i.e. less than C/10, C/10 is the ideal rate of charge) they can, after a while, achieve their fully charged voltage without achieving their full capacity.

Table 7.4 Characteristics of primary cells compared with miniature secondary cells

Type of cell	Size of cell	No. cycles	Nominal capacity (ah)	Useful[1] Energy (Wh)	Cell cost (US$)	Unit[2] cost (US$/kWh)
Zn-C	D	1	5.5	2.2	1.0	450.0
Alkaline	D	1	16.0	14.0	2.0	140.0
Ni-cad	D	100	4.0	400.0	8.5	21.0
	D	200	4.0	800.0	8.5	10.6
	D	500	4.0	2000.0	8.5	4.25

1 Useful energy (Wh) has been obtained from typical characteristic curves of dry cells and for a load typical for a small lighting application (e.g. a torch); discharge rate 0.5A (two cells in series for a 1.2W/2.5V bulb)

2 The unit cost indicates the cost of the battery per unit of output. A rechargeable battery also needs a charging source which adds to the cost a variable amount depending on the type of charger. Obviously the charger will generally last many more cycles than an individual ni-cad cell. Examples of 100, 200 and 500 cycles for the ni-cad cells are given.

In extreme cases ni-cad batteries may need an extended charge followed by discharge to a low level and then extended recharging with a higher than usual current (more than C/10), to 'exercise' the battery and attain the full capacity. This can be necessary after very long periods when the battery has been left discharged or when it may refuse to accept its full charge at C/10.

Cost-effectiveness of ni-cad batteries

The smaller sizes of ni-cad cells (sizes AAA, AA, C and D) have a higher initial cost than a primary cell, but work out much less expensive in the long run since they can be recharged and re-used from 100 to 1000 times before they lose their capacity and need to be replaced. Obviously a suitable power source is necessary to recharge them, which could be a special low-voltage charger (powered by the mains or a generating set) or solar photovoltaics (see Chapter 8 and Buyers' Guide).

Table 7.4 indicates the lifetime, capacity and hence the costs of primary cells compared with secondary cells when used in torches (flashlights).

Large nickel-cadmium batteries can also be financially competitive with large (over 100Ah) lead-acid batteries, bearing in mind that they can be 100% discharged while a lead-acid battery generally should be limited to 50-70% discharge of its rated capacity. Reference to Table 7.2 indicates battery costs averaged from data supplied by manufacturers.

7.4 Battery-based Systems

Battery chargers

The usual way to recharge either lead-acid or nickel cadmium batteries from a mains power supply is with a battery charger. Various types exist ranging from small units purpose-designed for recharging secondary batteries (i.e. sizes AAA, AA, C and D) to large units designed to deliver high-charging currents for the rapid recharging of large secondary batteries.

The smallest general-purpose chargers for standard torch or flashlight-sized ni-cads cost in the region of US$20 and typically take 12 hours or so to recharge four AA or two D size cells from the mains. Larger charging units can cost as much as US$300-$500 to deliver up to 20 amps in a carefully monitored and controlled charging cycle.

Battery-backed lighting systems

Some typical battery-backed systems are shown schematically in Figure 7.15. In all cases there is a battery bank which needs to be fed with DC electricity at the appropriate voltage to suit the batteries selected. In most cases multiples of 12V are common, especially 12 or 24V and in a few cases as much as 120V DC.

> CAUTION: It should be noted that DC voltages greater than 100V are potentially lethal.

Charging system

The power supply for recharging batteries can be either a DC source such as solar photovoltaic panels or it can be a transformed and rectified AC source (i.e. as supplied for mains appliances: 240V at 50Hz in most countries). AC can readily be changed from one voltage to another through the use of a transformer.

When a transformer decreases the voltage of an AC supply by a certain factor, the output current will increase by the same factor, therefore keeping the transformer input and output power more or less constant. In other words doubling the voltage halves the current and vice versa. For example if the current is 1A at 240V into a transformer that will step it down to 12V, the output will be a nominal 20A because 1A at 240V is 240W (i.e. 1A x 240V = 240W). Because the voltage is 20 times lower, the current is 20 times greater. In practice there is a small loss of power that appears as heat energy in the transformer, so slightly less than 20A will appear at the output. This illustrates an important point, namely that the lower the voltage the higher the current for a given electrical power demand.

After adjusting the voltage via a transformer the low voltage AC is usually transformed to DC using sold state electronic devices known as diodes. Normally the transforming and rectification of AC for battery charging are completed in a single 'black box'; this is the battery charger (sometimes known as a 'DC power supply'). This device commonly also includes some electronic control circuitry to sense the battery condition and hence to control the current to the battery.

Use of the electricity stored

The output from the battery can be used as it is to power DC lights, but generally this will limit the system for use with low voltage (usually 6, 12 or 24V) lamps. If numerous lamps are needed as part of a lighting network, then the use of DC low voltages will tend to require heavy and quite expensive cables because of the high currents, as can be seen from Table 7.5.

Table 7.5 illustrates that, for example, to transmit 10A at 24V (i.e. a mere 240W) requires 16mm² cross-section cable for a transmission distance of only 45 metres to keep the voltage drop to less than 10%. The

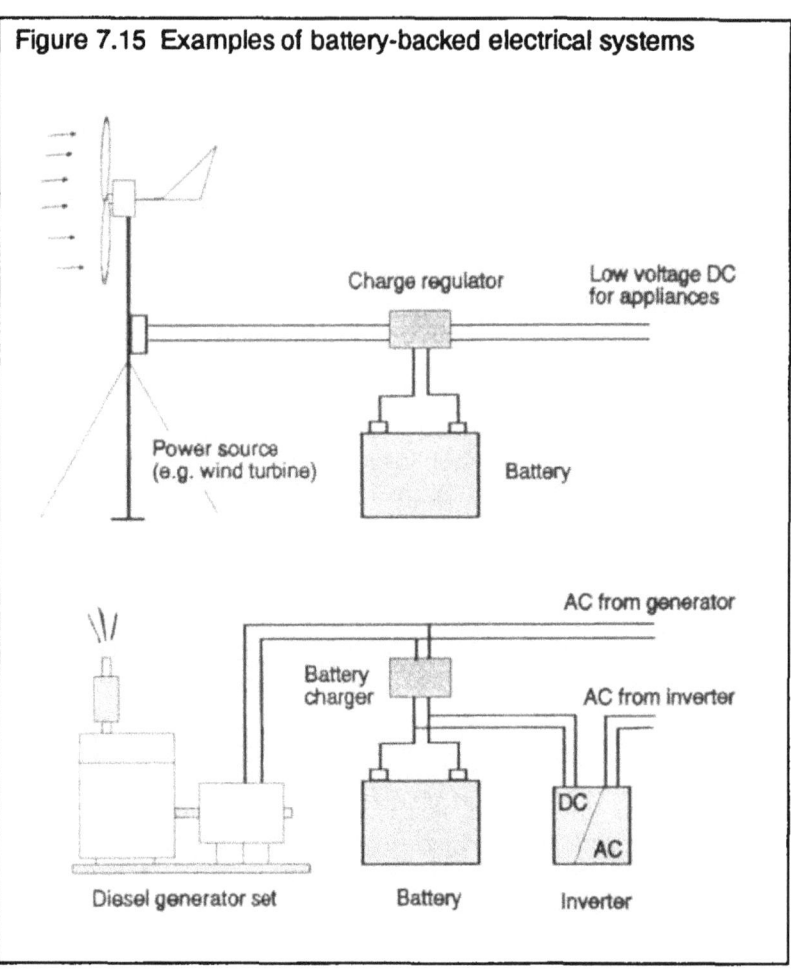

Figure 7.15 Examples of battery-backed electrical systems

percentage voltage drop gives a direct measure of the cable losses, so a 10% drop represents 90% transmission efficiency (i.e. 10% of the energy transmitted fails to arrive). In contrast, 240W at 240V requires only 1A so that even 1mm² conductor would allow 300m transmission distance with 240V before a 10% loss occurred.

Conductor (two core) mm²	Nominal rating A@240V	Nominal power kW @ 240V	Voltage drop per Amp per metre (mV)	Transmission distance for 10% voltage drop at 10A		Trade price US$/m
				@24V	@240V	
1.0	11	2.6	40	3	30	0.18
1.5	13	3.1	27	4	40	0.30
2.5	18	4.3	16	8	80	0.45
4.0	24	5.8	10	12	120	0.75
6.0	31	7.4	7	18	180	1.12
10.0	42	10.0	4	30	300	2.25
16.0	56	13.5	3	45	450	3.38

Table 7.5 Voltage drop and energy loss for different cable cross-sections

Figure 7.16
Battery charger used in a hybrid (diesel/solar) lighting system in the remote hospital of Kingoyi in Zaire. During the day, the batteries are charged by photovoltaic modules. Each evening, when the diesel generator of the village runs for three hours to power heavy loads (e.g. grain mill), some power is used by the battery charger (on the right) to top up the charge of the batteries if needed

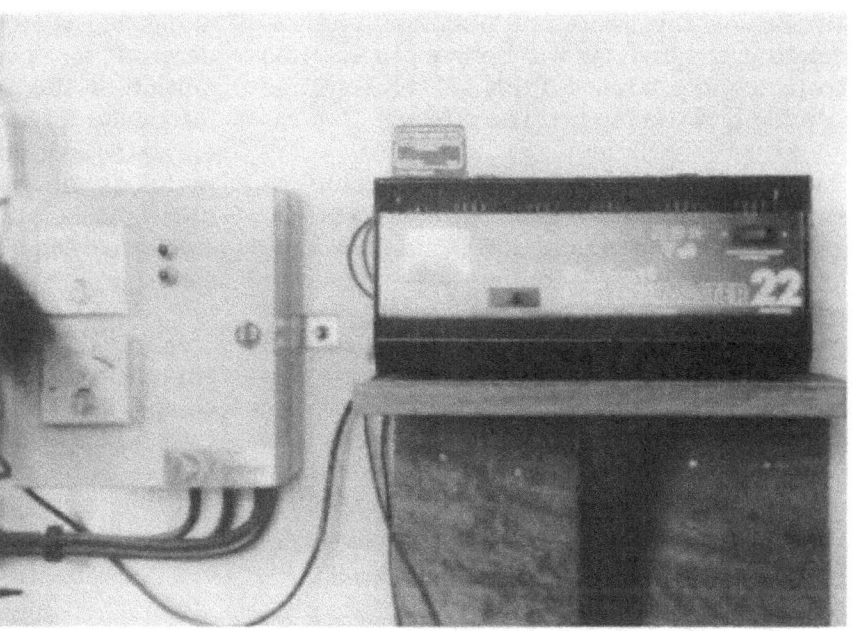

The point to remember is that the size of cable required depends only on the current and not on the voltage or power transmitted, so that higher working voltages mean that thinner cables can be used. A 12/24V system is therefore only feasible if transmission distances and currents are kept quite small, otherwise very heavy and expensive cables are needed.

Inverters

A solution to reduce power loss in cables is to convert the low voltage DC battery output back to AC mains voltage. Not only does this allow a less expensive cable system but it also allows the use of unmodified standard mains lights (and other mains electrical appliances) which tend to be easier to purchase and less expensive (but usually less energy efficient too) than special low-voltage units.

The conversion from DC to AC is usually done by using a device known as an inverter which accepts low voltage DC and outputs higher voltage AC. Inverters are generally quite expensive and the specifications vary widely. The cheapest and crudest inverters produce a square wave AC output and are relatively inefficient, even at rated power, and often highly inefficient at part load.

More sophisticated inverters cost more, but produce a 'quasi-sinewave' output (i.e. approximately equivalent to normal mains AC power). They tend to be more efficient at rated

power (the best are around 90% efficient) and a lot more efficient at part loads. The very best are designed to switch themselves off at no load. A danger with inverters that do not have this feature is that they can drain the batteries to which they are connected due to their own parasitic power demand even when no power is being drawn.

Small inverters are relatively new devices in the consumer market and until recently they had a tendency to be unreliable. It is certainly worth trying to find an inverter that has a 'track record' and is widely used, as well as having good efficiency right across its power spectrum from 'no load' to 'maximum rated power'.

Another important feature with inverters is their overload capacity. The better models can handle from four to six times their rated power for brief periods. Since many electrical appliances (especially devices with electric motors such as power tools or refrigerators) draw a surge of current greater than their normal power requirement when first switched on, some overload capacity is generally an important requirement. Otherwise, there is a risk of burning out either the inverter or the appliance.

Inverters have a tendency to create radio frequency interference that can cause noise when listening to a radio. This can often be suppressed with special add-on electronic components. Again, it is worth questioning a supplier on this issue and obtaining guarantees that the inverter will be free of any radio frequency interference if this is likely to cause a problem in the intended area of use.

The most sophisticated inverters come as combined inverters and battery chargers. They effectively act as a UPS (Uninterruptible Power Supply) when connected to a battery bank. When power is received by the inverter-charger unit (whether from an unreliable mains supply or from a diesel generating set) and if the battery voltage is below a pre-set value the unit will charge the batteries and connect the mains or the diesel power supply to the load. Once the battery voltage reaches the level indicating they are fully charged the inverter-charger ceases to charge. If the

mains/diesel stops and there is a continuing demand for power from the 240V 50Hz output of the inverter-charger unit, it immediately switches in the battery as an energy source and acts as an inverter. If the mains or the generator comes back on, the unit links it straight back to the load and starts to recharge the battery again at the same time.

Some inverter/charger units have a sensor to cut off the system if the battery state of charge falls below some minimum pre-set level, to prevent over-discharging and hence battery damage. In this case no power will be available until the diesel or the mains come back on, but at least the valuable batteries will be safeguarded.

Example calculation for sizing a battery

Requirement: to run 300W of lights (e.g. 15 CFL lamps of 20W/240V AC each) for 6h per night plus a 100W/240V AC TV for 5h per night. The daily energy requirement will be 300W x 6h + 100W x 5h = 2300Wh.

Since an inverter will be required to convert battery power (24V DC) to mains power standards (240V AC), we must assume some losses from the inverter and cabling. An efficiency of 70% might be a reasonable guess for a reasonably modern and well-cabled system, so the gross output from the battery will need to be:

$$2300Wh / (70/100) = 3285Wh \text{ every } 24h$$

If the system voltage is 24V DC, the battery **usable capacity** is:

$$3285Wh / 24V = 137Ah$$

If we assume a depth of discharge of 50%, we will need a battery of:

$$137 / (50 / 100) = 274Ah \text{ at } 24V$$

This gives an idea of the minimum battery size to look for.

Cabling: If six batteries of 100Ah at C/20 and 12V are chosen, the batteries will be connected as pairs in series (to give 24V) and the three series pairs connected in parallel to give the required capacity. Consequently, each battery will supply one third of the total current required. The maximum peak current given to the load is 16.6A (i.e. (300W + 100W) / 24V = 16.6A). Each series pair will supply 16.6/3 = 5.5A, which is just above the C/20 rate of discharge for the rated capacity. Therefore the battery choice is correct.

NOTES

1 The size of the battery will allow only one day of autonomy so this assumes that the battery bank is being recharged every day after use. If not, the size of the battery bank has to be multiplied by the number of days without possible recharging (e.g. in solar system, estimated number of consecutive days without sunshine).

2 If the required appliances are available at 24V DC, there is no need for an inverter. Therefore the battery size could be reduced by 30%, but the cables would need to be thicker.

Choosing and sizing a battery

The sizing of a battery will determine its performance and lifetime. A rule of thumb to size a battery for a lighting system is illustrated by the following example.

Selection and purchase of a battery

The characteristics required for a battery system to perform well are:

- High cycle life;
- Low maintenance requirements;
- Reliability;
- Low rate of self-discharge;
- High discharge/charge efficiency;
- Ability to survive complete discharge.

When selecting and buying a battery, it is advisable to:

- Talk to people who have used batteries in similar applications and under similar conditions;

- Contact battery suppliers and;

  Give the following information:
  - ✓ use of the battery (how many amperes for how many hours);
  - ✓ voltage needed (inverter or appliance input voltage);
  - ✓ temperature (ambient temperature)
  - ✓ cycle length (typical charge-discharge cycle period);
  - ✓ life expected (how long the battery should last).

This information should enable a manufacturer to recommend the most cost-effective solution to meet your duty requirement.

  Ask for the following information:
  - ✓ type of battery with alloy contents;
  - ✓ cycle life at several values of DOD and temperature;
  - ✓ variation of capacity with rate of discharge and cut-off voltage;
  - ✓ rate of self-discharge;
  - ✓ relation between specific gravity of the electrolyte and the state of charge;
  - ✓ yearly distilled water consumption;
  - ✓ complete installation, operating and maintenance instructions;
  - ✓ warranty.

However, manufacturers' warranties are of limited use in remote areas as the cost of returning a faulty battery for exchange may be prohibitive and few warranties will cover shipping costs. They are nevertheless worth reviewing, if only as an indication of the confidence a manufacturer places in their own product.

- Compare your sizing with the manufacturer's own sizing. If large differences exist and remain obscure, discuss them with the manufacturer.

- If possible, verify manufacturers' claims by testing a sample of batteries under the simulated service conditions.

Influence of users

Last but not least, the life of most batteries depends greatly on their use, and hence on the users. It is of great importance that users are trained to use and maintain the batteries for their best performance.

Figure 7.17
Maintenance of a battery; the technician is applying vaseline to the battery terminals to avoid corrosion. Note some of the necessary tools and spares needed for the maintenance of a battery; hydrometer; multimeter; thermometer; wooden gauge to measure the level of electrolyte; spanners; distilled water and cleaning cloths

Electricity Generation 8

8.1 Introduction

It has been emphasized throughout this book that electricity is the preferred way to provide power for lighting. However, it has been estimated that at present, in the developing countries, only one rural inhabitant in five currently has the use of mains electricity. This 'non-electrified' population amounts to about half of humanity, well over two billion people. Rural electrification is therefore a major problem for the human race as a whole, even though it is a problem that is rarely considered by the 'electrified' urban populations.

Although extending the grid is one option, in practice this is frequently not feasible either technically or economically, and so other methods of small-scale localized electricity generation need to be considered. These will be reviewed in the following chapter.

able in some places and not in others. Similarly, the quality of the grid also varies. In many remote areas the grid can be quite unreliable with frequent power failures or 'brown outs' (i.e. extended periods of supply with the voltage significantly below normally specified levels). This is particularly true in countries where the national power system is already under strain to meet the existing demand for power.

The possibility of using secondary batteries, which can be recharged from the mains, for lighting is one method of ensuring more reliable lighting in situations where the grid supply frequently fails or is shut down for extended periods.

One of the problems of extending the grid is that most areas currently not served by it tend to have a low population density which

8.2 Grid Extension

If a reliable mains power supply is located not far away, there are two options by which it may be used to provide light. These are either;

- Extending the mains to enable it to be used directly for lighting, or;

- Transporting secondary cells or rechargeable batteries to the nearest place with a mains supply for regular re-charging.

The cost of grid extension varies and subsidies are avail-

Figure 8.1 The lighting and entertainment system of this bar in Zaire where there is no mains electricity, is provided by a battery recharged weekly at a town 35km away

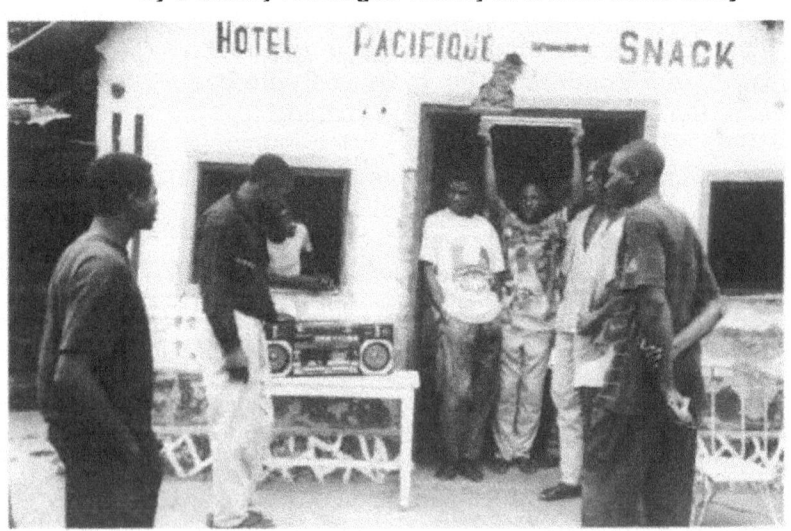

implies a need for a greater amount of connection cable per person connected, and hence higher than average costs. Furthermore, the demand for electricity from remote rural populations will tend to be lower than from urban residents, thereby leading to lower levels of revenue to cover the higher costs.

Moreover the line losses (i.e. unpaid-for power loss) are invariably greater for a rural power distribution network than for an urban one due to the combination of long cables and low power factors. Therefore, unless subsidies are provided for rural electrification, the true costs of delivering mains power to rural people can be very much higher, per capita, than for the urban population.

Table 8.1 indicates some typical village electrification costs (in 1985 US$).

Cable and material costs, alone, amount to between US$1500 to $5000 per kilometre, depending on the cable cross-section specification and whether armoured cable is buried in the ground or bare cables are used on insulated poles. The costs of shipping, labour, administration and transformers, and other connection interfaces need to be added to this amount as an overhead. Therefore, typical low-voltage grid extension costs are in the region of US$5000 to $15,000 per kilometre at the time of writing.

It follows that, unless a house or settlement is within a few hundred metres of an existing grid network, any extension, especially if it is 'only' for lighting (which most of the time has no direct economic value unless it is for the lighting of a workshop or a bar at night) is difficult to justify in economic terms.

If the grid (or a regularly operated diesel generator system) is within a few kilometres, a feasible option which is quite widely practised, is to use rechargeable batteries for small-scale lighting. Normally two batteries are needed; one in use at any one time and the other away being charged. Then, particularly if regular transport is available, the two batteries may be swapped every few days to allow the discharged one to be recharged.

A common 'mistake' is to use a car battery for this purpose. Since car batteries will not

Table 8.1	Typical village electrification costs (in 1985 US$)
Country	Mean cost per village
India	25 000
Bangladesh	70 000
Zambia	73 000
Tunisia	83 000
Bolivia	105 000
Pakistan	95 000
Indonesia	125 000
Benin	300 000
Burkina Faso	320 000

Notes

1. Source - Background Paper, UNDTCD/AFME Symposium on Economics of Small Renewable Energy Systems for Developing Countries, Sophia Antipolis, France, June 1986.

2. In India, which produces most of its own equipment, the cost per km of low-voltage power line was US$2100, whilst in Benin, where everything has to be imported, it was US$16,000.

survive any great depth of discharge for more than 20 cycles or so (as explained in Chapter 7) they tend to have a short useful life. A battery recharging and swapping system will therefore work much better if low-antimony deep-discharge storage batteries are used, although these batteries are often difficult to obtain in remote areas.

In practice, in developing countries, people can rarely afford to buy either two sets of batteries or deep discharge batteries. In recent years, in some African countries (e.g. Morocco and Kenya), automotive battery manufacturers have been producing a so-called 'improved car battery' to fill the need for a deep-discharge battery. These batteries have thicker plates with a low antimony content (2 to 3%) and usually a larger electrolyte reservoir than automotive batteries. They are slightly more expensive than automotive batteries but last longer.

8.3 Small Generator Sets

Internal combustion engines can be divided into two main categories, petrol, gasoline or kerosene-fuelled spark ignition (s.i.) engines and diesel-oil-fuelled compression ignition (c.i.) engines.

Generator sets are supplied with alternators to produce 240V or 220V, 50Hz (or 120V 60Hz) to suit standard mains lighting (or other) appliances, although smaller units can be obtained to deliver 12 or 24V DC.

Most applications that involve no more than lighting only need a few hundred watts at most, so the only practical generator set for such purposes is a petrol, gasoline or kerosene spark-ignition engine.

Small engines are usually derated to about 70-80% of their rated power. For instance, a 5kW rated engine will be necessary to produce a continuous power output of 3.5-4.0kW at its shaft. In addition, because there are losses in a generator (a small unit may only be 80% efficient) the electrical output from such an engine when used in a generator set would be 80% of 3.5 to 4.0kW or 2.8 to 3.2kW, which is only 60% or so of the engine shaft rated power.

The main reason for derating an engine is to prevent premature wear. In addition the optimum efficiency for most engines is achieved at a speed corresponding to about 70 or 80% of its speed for maximum power. Therefore, derating an engine usually improves its fuel consumption. Further derating is necessary at high altitudes or at high ambient temperatures, and recommendations

to this effect are usually made by the manufacturer.

Petrol (gasoline) engines

Petrol engines are manufactured with minimum power ratings of around 300W, delivering electricity as either 12 or 24V DC or 120 or 240V AC. The largest s.i. sets are generally around 5kW. Table 8.2 gives characteristics of a range of typical portable small petrol (gasoline) engine generator sets.

Figure 8.2 Example of a petrol generator set (2kW / 220V AC, overall size 0.6m x 0.5m x 0.4m, 45kg)

The efficiencies given are based on typical manufacturer's fuel consumption figures, which almost always apply to a new engine running under ideal conditions. In reality however, the fuel consumption for small engines can be as much as 50 to 100% higher (i.e. 1.5 to 2 times the consumption indicated). However this does indicate clearly how the smallest size of engine tends to be much less efficient than those only slightly larger.

Small spark-ignition engines wear quite quickly as they have small components and usually run at quite high speeds. The smallest engines will probably need a major

Table 8.2 Typical characteristics for three sizes of SI generator sets			
Rated output (shaft) (kW)	0.6	2.8	4.0
Rated output (electrical) (kW)	0.4	1.0	2.0
Dry weight (kg)	20.0	38.0	44.0
Fuel consumption (l/h)	0.5	0.7	0.9
Efficiency (%)	9.0	16.0	24.0

Figure 8.3 A 600W/110V petrol generator set (overall size 0.4m x 0.3m x 0.35m, 20kg)

overhaul with replacement of key components after about 1000 hours continuous use and the slightly larger ones might manage up to about 2000 or 2500 hours between major overhauls. At three hours per day, 1000 hours is reached after only 330 days of use, which is between 10 and 11 months.

The total operational life for such an engine will be in the region of 2500 to 7500 operating hours (typically one to five years between engine replacements).

Although s.i. engines are lighter, more portable, cheaper and in some ways easier to maintain than diesels, they are also less efficient and use much more fuel than a diesel. They also tend to be less robust and to have an inherently short operational life, making them generally unsuitable for more than two or three hours use per day. However, small spark-ignition engines are much quieter than diesels and can be obtained as sound-proofed portable units. They are increasingly popular in this form for powering street-traders' (hawkers) stalls in the cities and towns of South-East Asia, where the few hundred watts of electricity are used for a combination of lighting and powering a tape player which usually 'blasts' out music loudly enough to drown out the sound made by the engine.

Fuel for an s.i. generator set: Gasoline

Gasoline or petrol is a mixture of the most volatile liquid hydrocarbons obtained from refining crude petroleum. Safe storage of such a volatile fuel is difficult and gasoline vapour can easily escape from any container and constitute a permanent potential source of fire and explosion in houses or shops.

The price per litre of gasoline can be quite high. For instance, in inland Sahelian countries of West Africa, gasoline costs around US$0.85 to $0.20/litre in the capital cities and often as much as US$2.0/litre in rural areas.

Diesel generator sets

Diesel generator sets are manufactured in sizes from a few kW up to several megawatts. The smallest diesel generator sets are around 1.5 to 2kW and more commonly around 5kW. Some characteristics of small diesel generator sets are indicated in Table 8.3. Since diesels tend to be damaged by extended operation at less than about 25% of their rated power, they are not very suitable for small applications involving only lighting.

The comments made about the performance figures of petrol engines apply equally well to the performance figures given here, except diesels tend to be more efficient over a broader band of operating conditions than petrol engines. Therefore, the 'real' fuel consumption is more likely to be in the region of 25-50% worse (i.e. 1.25 to 1.5 times more fuel per hour may be needed in practice than the optimistic manufacturers' figures given).

Table 8.3	Typical characteristics of three sizes of small diesel generator sets		
Rated output (shaft) (kW)	2.5	4.5	9.5
Rated output (elect) (kW)	1.5	3.6	7.1
Dry weight (kg)	150.0	290.0	390.0
Fuel consumption (l/h)	0.7	1.5	2.7
Efficiency (%)	20.0	22.0	25.0

Fuels for diesel generator sets: Gas oil/diesel fuels

Oils distilled from petroleum from about 110°C upwards are called gas oils because originally they were used to produce illuminating gas by a cracking process. Diesel fuels can be classified from light gas oils, which are mainly used for small diesel engines (e.g. car and small generator set engines) to medium fuel oils which are used for large low-speed engines (e.g. boat engines).

Diesel fuel or gas oil costs much less than gasoline in most countries because it is usually untaxed. It is commonly available only at petrol stations in cities and towns, or along main roads. The price per litre sometimes increases with the remoteness of the location.

Battery-backed diesel systems for lighting purposes

Diesel engines are usually used directly for lighting power supply after dark but are often sized to power much larger loads during the daytime (e.g. a workshop). A problem that can occur therefore, is that the lighting load alone tends to be too small for the diesel. Running a diesel for long periods with less than 25% of its rated load can cause carbonization of the injectors and gumming of the piston rings. Therefore many people who use diesel systems deliberately leave electrical appliances turned on when they are not really needed just to maintain the load on the diesel. It is common practice to leave lights on all day with a small diesel generator set if it is not being effectively loaded.

Also, because reducing the load on a diesel engine does not save a proportional amount of fuel, there is little incentive to use high-efficiency energy-saving lights such as fluorescent tubes, compact fluorescent lamps, or exterior low-pressure sodium lighting, since the reduced load on the diesel which they offer could be more of a problem than a help at times. The problem is that a diesel

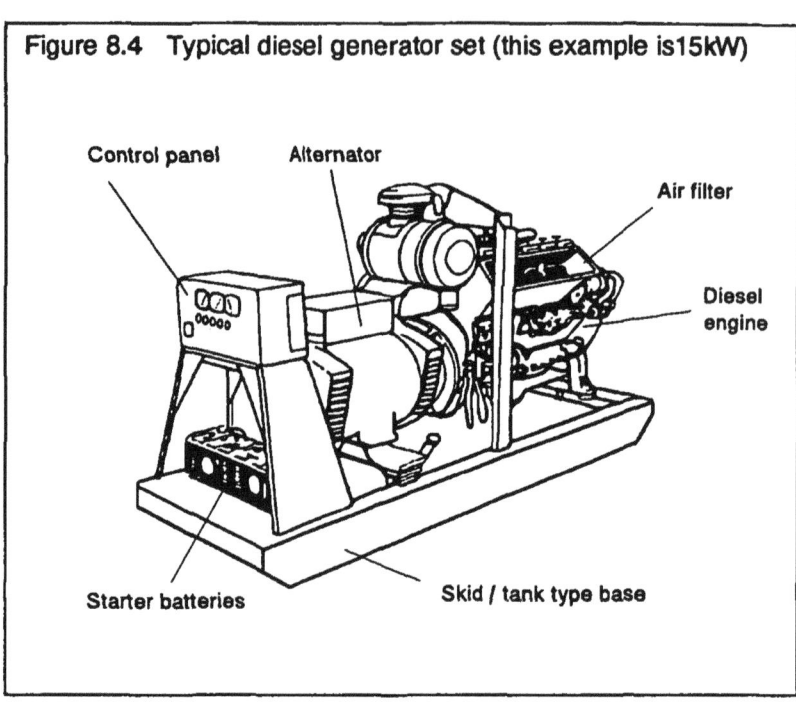

Figure 8.4   Typical diesel generator set (this example is15kW)

Control panel   Alternator   Air filter   Diesel engine   Starter batteries   Skid / tank type base

generator set is sized to meet the peak daytime demand which is then too large for the evening 'base load' consisting usually of just a few lights.

In conclusion, the best way to cut diesel running costs is to run the diesel for fewer hours per day, but at full load. This sometimes becomes possible if the 'key' lights (e.g. living room, bedrooms and bathroom) are run from batteries, which can be recharged from the diesel during the daytime. If this arrangement is used, then in many cases the diesel need not be run at all at night. Moreover, such an arrangement has the further major 'operational advantage' that the key lights will be available at any time of the day or night, at the flick of a switch, regardless of whether the diesel is running, (providing of course that there is an adequate state of charge in the batteries).

Since batteries are expensive items, there is of course a need to minimize the load supplied by them to keep costs down. Therefore once a system of this kind is in use there is every incentive to switch off any lights that are not needed at times when the diesel is not running and to use only the most efficient types of lamp such as fluorescent tubes. Systems of this kind could be expanded to a sufficient size to power other relatively low-powered domestic appliances such as a TV or a refrigerator.

8.4 Solar Photovoltaic Systems

Solar photovoltaic cells are solid-state electronic devices that convert sunlight directly to electricity without any moving mechanical components. Solar photovoltaic (PV) technology is therefore extremely reliable, long-lasting and has almost no running costs and only very low maintenance requirements. It is therefore one of the most promising power sources for small-scale lighting applications off the grid and a growing range of solar-powered lamps and lighting kits are emerging on the world market (see Buyers' Guide).

Crystalline silicon modules

Figure 8.5 illustrates a solar photovoltaic module, which is the basic power source for solar electricity. The most common type consists of a number of thin discs (known as cells) of crystalline silicon mounted behind glass in a sealed, weather-proof and dust-proof enclosure. The surface of the silicon cells is treated in various ways that make it respond to sunlight by developing a voltage between the front illuminated face and the back surface. A network of fine metal conductors is applied to the front of each cell and the back is metallized all over. The conductors on the face of one cell are then connected to the back of the next cell and so on. Each cell produces about 1W with an open circuit-voltage of 0.5-0.6V. By linking them in series a complete module can produce a voltage suitable for charging a battery. Typically 36 cells are connected in series to achieve an open-circuit voltage of 16 to 18V, which will have an operating voltage of 15V, suitable for charging a nominal 12V battery.

The most modern types of module with separate cells wired together are based on square cells (e.g. 100mm x 100mm) made from either poly-crystalline or mono-crystalline silicon. The latter are more efficient but more expen-

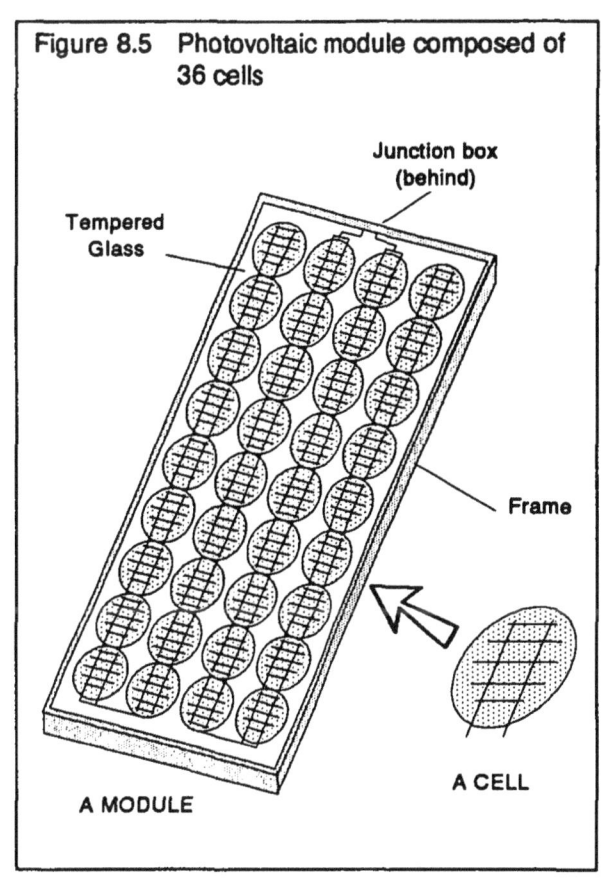

Figure 8.5 Photovoltaic module composed of 36 cells

Junction box (behind)

Tempered Glass

Frame

A CELL

A MODULE

sive. Mono-crystalline modules are also sometimes made of circular cells. Mono-crystalline cells have a uniform blue colour while poly-crystalline are also blue but display a pattern similar to that on a piece of galvanized metal.

Figure 8.6 Different sizes of solar modules

Typical light-to-electricity conversion efficiency is in the range of 12 to 14% (i.e. one square metre of modules can produce 120 to 140W at noon). A typical module is 0.4m x 1.0m in size and may cost in the region of US$300 to $400.

When modules are connected together, they are called an array. Since in some applications, especially small portable lamps, a typical module is much larger (and more expensive) than necessary, smaller modules are manufactured which use half-cells or quarter-cells to achieve the same voltage as the full-size modules but with a quarter or half the power under given solar conditions.

Thin film technology

Modules can also be fabricated from a continuous thin film of amorphous silicon deposited onto the back of the glass window. These are less efficient but cheaper than crystalline silicon modules. Ultimately the cost per unit of power is fairly similar between different types of module at the present time. The main disadvantage of thin film technology is that performance decreases with time, with efficiency eventually stabilizing at 4 to 6%. Due to this low efficiency, the size of a module with a given power rating needs to be at least

double that of a corresponding crystalline module. Amorphous silicon modules are commonly used for low-power applications (e.g. up to 20W.)

There are also other types of solar cell (e.g. cadmium telluride or multi-junction types) but they are not yet commercially available and it is beyond the scope of this book to give details of these. The interested reader is referred to *Solar Photovoltaic Products: A Guide for Development Workers* which is in the same series as this book and which gives much more detailed information on all aspects of photovoltaic technology.

Energy from solar photovoltaic modules

The standard crystalline silicon (0.4m x 1.0m) PV module delivers 40 to 50W at 12V given 1000W/m² sunlight (i.e. strong midday sunshine).

It is usual to position a photovoltaic module or array to face south in the northern hemisphere and north in the southern hemisphere. Ideally, to gain the maximum energy, the photovoltaic module can be mounted on a tracker so as to face the sun continuously, but this is not practical in most cases so the

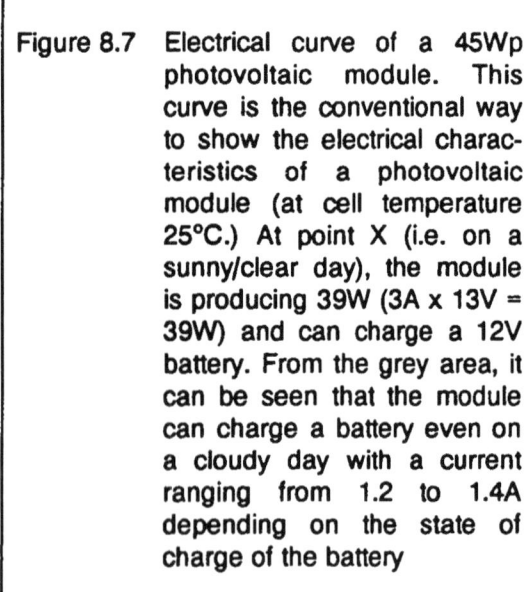

Figure 8.7 Electrical curve of a 45Wp photovoltaic module. This curve is the conventional way to show the electrical characteristics of a photovoltaic module (at cell temperature 25°C.) At point X (i.e. on a sunny/clear day), the module is producing 39W (3A x 13V = 39W) and can charge a 12V battery. From the grey area, it can be seen that the module can charge a battery even on a cloudy day with a current ranging from 1.2 to 1.4A depending on the state of charge of the battery

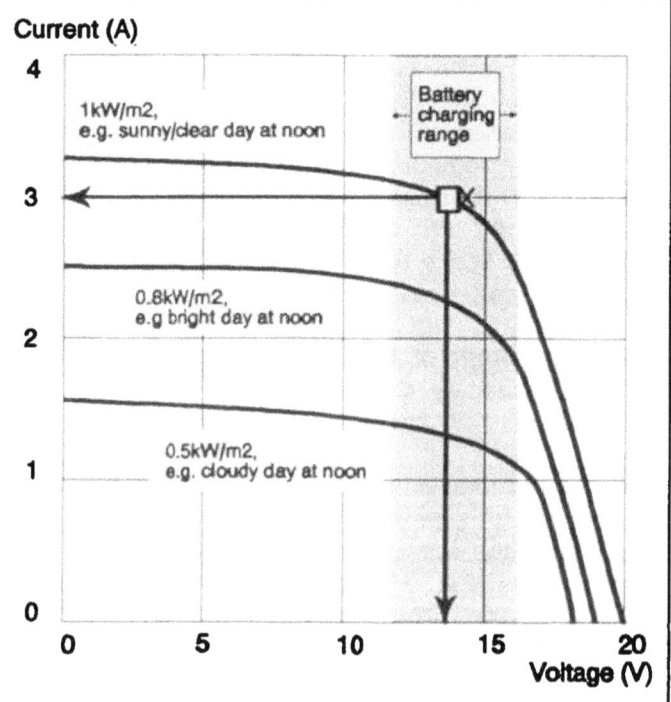

best compromise for a fixed solar panel is to face square to the sun at noon. If the module is inclined from the horizontal surface of the ground at a 'tilt angle' corresponding to the latitude of the location, then it will be square to the sun at noon at the Equinoxes (March 21 and September 21). A steeper angle increases the output in winter; a shallower angle gives more output in summer.

In the equatorial and tropical regions, the tilt angle should be at least 10° in order to ensure good runoff of rain (to keep the panel clean) and also for better cooling.

For a small lighting system, a manually operated tracking module support may be an option, as it may increase the output by up to 25% if the user is operating it properly. Therefore the array can be correspondingly smaller. However if not attended to regularly to adjust the position, the battery will suffer deep discharge leading to its premature failure. In conclusion, unless the user is well trained and highly motivated, it is not recommended to use movable module supports. A further disadvantage is that tracking supports are more susceptible to storm damage.

A photovoltaic module will give substantially more power if cooled. It is advised to allow a free air circulation all around the module. For example, if the module is installed on a roof, a

space of at least 10cm should exist between the roof and the back of the modules allowing air to cool the modules.

Sizing

Modules are rated in terms of peak watts (Wp). This is the power it will produce in full sunshine, defined as 1000W/m². This is about equivalent to the intensity of the sun at noon. Hence a 40Wp module will produce a maximum of about 40W in full sunshine.

Accurate sizing of a solar panel is normally done by the system supplier. For those who wish to design their own system and assemble it from correctly sized components, the topic is dealt with in more detail in the publication *Solar Photovoltaic Products* mentioned earlier.

However here is a quick way to estimate the daily energy output of a module. The maximum energy produced by a module can be approximately calculated by multiplying the power rating by the average insolation over a typical day:

Module energy output (Wh) = Module power rating (Wp) x Insolation (kWh / day / m2)

This formula can be illustrated by means of the following example. If the incident energy

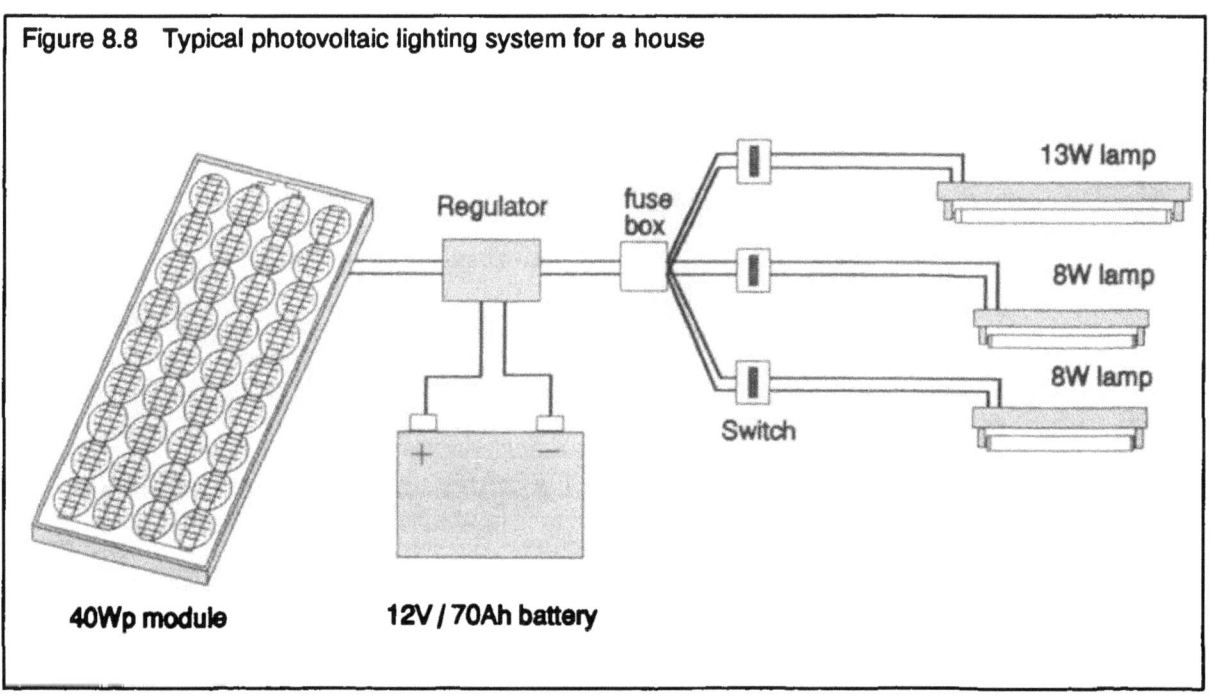

Figure 8.8   Typical photovoltaic lighting system for a house

Regulator

fuse box

13W lamp

8W lamp

8W lamp

Switch

40Wp module   **12V / 70Ah battery**

from the sun was 5kWh/m² summed over the whole day (e.g. in regions ranging from moderately sunny to very sunny), this is the same total energy as if the sun stayed at noon giving 1000W/m² for five hours (1000W/m² x 5 = 5000Wh/m²). As a module rated at 50Wp gives 50W when receiving 1000W/m², it will produce 250Wh over the whole day when exposed to the equivalent of five full-sun hours (50Wp x 5kWh/m²)

If the daily insolation is only 3kWh/m², a 50Wp module will produce 150Wh/day (i.e. 50Wp x 3kWh/m²).

Since batteries are usually about 70% efficient, a single 50Wp module and an appropriately sized battery will yield an average of 105 to 175Wh of useful electrical energy for lighting in a solar regime ranging from 3 to 5kWh/m² per day.

Taking the lower figure to allow for cloudy conditions even in a sunny climate, and adding a margin of safety of 50%, implies that a single 50Wp module can typically provide

Figure 8.9 Solar portable lamp with a 7.5Wp amorphous silicon module (Courtesy of Neste/Naps)

50Wh per 24 hours period which is enough to run a 6W fluorescent tube for eight hours. Therefore a typical single module, even with a fairly pessimistic estimate, can sustain a reasonable size of lamp for several hours per night.

Solar photovoltaic lighting technology

The main difference between solar photovoltaics and other electrical power sources for lighting is that solar cells produce DC (direct current) electricity at low voltages. This is ideal for charging batteries which are an essential system component to store solar-derived energy for use at night. Therefore DC lighting units are the preferred option, although standard AC 'mains' lights can be used if an inverter is included in the system. However, the inverter efficiency will probably be no better than 50 to 90%, so a larger and more expensive solar array will be needed to make up for these losses. Moreover, an inverter can also be quite expensive and introduces another potential source of failure in the system.

The most cost-effective lighting option to use with a solar photovoltaic system is low-voltage fluorescent lighting or in some cases high-efficiency halogen lamps or sodium lamps (the latter for exterior purposes only).

Solar photovoltaic lighting applications

PV lighting systems can be categorized into three distinct groups:

● Portable solar lamps;

● Domestic or community building lighting systems (homes, schools, community centres, shops, places of worship, etc.);

● Other lighting applications: street lights, security lighting, etc.

Portable lamps

These are a single unit with a lamp, battery and a small solar photovoltaic panel all integrated. Sometimes the solar panel is detachable so that it can be located a few metres from the lamp, which can be useful from the point of view of positioning it in a window or somewhere where it will be well exposed to the sun during the day (Figure 8.9). This also prevents the battery from being exposed to heat and sunlight, both of which can lead to premature failure.

A dedicated portable solar lamp or lantern with a separate module usually comes complete and ready to use and merely needs to have its array left in a well-illuminated place during the

Figure 8.10 Solar torches with integrated PV modules. These torches can also charge ni-cad batteries which can be used to power either the torch or small radios, clocks, etc. The torch in the middle is equipped with a 4W fluorescent tube

day to ensure that the battery gets well charged. The solar module will usually be a non-standard miniature type (often with quarter cells) and may be in the range from 5-10Wp rating. The lamp will be a 4 to 8W fluorescent allowing operation for three to five hours per night. Typical costs are US$100 to $300 depending on capacity and specification.

Recently, some manufacturers have introduced solar hand-torches (flashlights). These are the size of a typical small hand-torch and are fitted with integrated PV cells able to recharge size AA nickel-cadmium batteries (see Figure 8.10). The light output is quite similar to torches powered by dry batteries and their purchase cost is in the range of US$20 to $40 including rechargeable batteries. Millions of torches powered by dry batteries used today by villagers in developing countries could be replaced by the solar torches. This would be more cost-effective for the users and would avoid pollution by the uncontrolled disposal of small dry batteries.

Domestic or community building lighting systems

Where more than one light is involved it is usual to use a suitably sized fixed solar photovoltaic array. The solar array can often

be conveniently located on the roof of the building to be illuminated.

Such systems usually include:

- PV module or appropriately sized array (several modules connected together) including mounting structure, cables and connectors;

- Fluorescent lights (typically rated at 8, 13, 16 or 20W) or high-efficiency low-voltage halogen lamps;

- One or more batteries;

- Charge control regulator;

- Any special tools, spare parts, plus comprehensive installation and maintenance instructions.

A 50Wp module system and lamps will cost from US$400 to $600. This may sound like a high price. It should, however, be compared with the cost of a grid connection which costs thousands of dollars per kilometre, or with the continuous running costs of procuring kerosene for a pressure lamp, which is the only form of non-electric light that is comparable in quality, if not controllability. Figure

8.8 shows a typical photovoltaic module wired to a control unit and a battery that forms the basis of a PV lighting system.

Other lighting applications

PV-powered street lights come complete with a mounting pole and array structure (see Figure 8.11). They typically include an automatic sensor to switch the lamp on at dusk and to switch it off again after a preset number of hours or if the battery state of charge falls too low. Street PV lighting systems are usually equipped with low-pressure sodium lamps or with compact fluorescent tubes because of their high luminous efficacy.

Security lighting can be provided with infrared sensors that only switch the light on if it is dark and some movement of a person, motor vehicle or large animal is detected by the sensor. Such a device is useful as it allows quite a powerful lamp to be used to illuminate a large area, but it only switches on when it is needed so as to minimize the energy demand and hence the system costs. Costs vary considerably depending on specification. A typical small security or street light installation with two solar modules (80 to 100Wp total) would retail for around US$1600 to $3000.

Summary of photovoltaic lighting features

In conclusion, it is worth summarizing the undoubted advantages and few disadvantages of solar photovoltaic lighting, especially as this is one of the most promising methods for providing good-quality controllable lighting for the future in areas lacking a grid electricity supply:

- PV systems need no fuel supply;

- PV systems are modular (so the system can readily be expanded to meet an increasing need for more light);

- PV systems are highly reliable: the lack of moving parts makes them significantly more reliable as an electrical power source than almost any other small-scale method for electricity generation;

Figure 8.11 Photovoltaic street lighting

- PV systems are easy to maintain and repair: there is little more to do than occasionally to clean the array. Faults are rare and usually are not difficult to trace and rectify;

- PV modules have a long life: crystalline silicon cells have been found to show little degradation in performance even after more than 20 years continuous use;

- PV power is environmentally benign; there is no pollution or hazard to the environment from the use of solar PV power, nor is there any heat or noise to cause local discomfort;

- PV lighting can be the least-cost method for providing small levels of reliable electric light on a decentralized basis, even though initial costs may be high;

- The main disadvantage of solar photovoltaic lighting systems is their high capital cost. However, costs have fallen significantly over the last 10 years and there is some hope that costs will continue to fall in the future, through further technical development.

8.5 Wind Electricity Generators

The use of small wind turbines to generate electricity will, in some cases, provide an alternative to the use of solar photovoltaic power. Wind is an intermittent energy resource, like solar energy, which consequently needs to be used to charge a battery if power for lighting is to be available at any time, including during calm weather.

The main regions where the use of wind power will be of interest are not so much in the tropics but in the higher latitudes where long winter nights and short winter days make the demand for energy to run lights greatest in the winter, when winds tend to be strongest and there is little available solar energy. Small wind turbines also work well in many coastal regions or on mountains and in areas with harsh winter climates where such factors as salt spray (from the sea), dust or snow cover can make the reliable use of solar photovoltaics problematic at certain times.

Wind electricity generators for lighting systems

The main type of wind turbine of interest for lighting purposes will be a small machine charging a battery for powering the lights.

Figure 8.13 illustrates a typical 50W wind turbine with a 0.9m rotor diameter. Windturbines intended for battery charging are marketed with power ratings of 20W, 50W, 100W, 250W, and in various ratings right up to 10kW from various manufacturers. It should be noted that this power rating is defined for each turbine at its 'rated' windspeed (usually around 10 to 12m/s for small machines) which is usually far above the average windspeeds that might be expected in normal use.

However, systems of more than a few hundred watts are complex and since they will generally be used for other purposes in addition to light-

Figure 8.12 Thousands of small battery-backed lighting systems charged by small wind generators have been installed for lighting of the nomads' yurts in Mongolia

ing, this book will only focus on the use of the smaller types of wind electricity generator purely for charging a battery for lighting purposes (and for other small electrical applications like radio and TV).

These machines generally consist of a small rotor, often using glass-fibre-reinforced plastic blades, which drive a permanent magnet alternator. The better designs have low-speed alternators directly driven by the rotor. More old-fashioned designs use DC generators sometimes driven at higher speed via a

Figure 8.13 A 50W wind generator designed specifically to charge 12V lead-acid batteries

Figure 8.14 12V/11W folded fluorescent lamp installed inside a yurt in Mongolia

gearbox. DC generators and gearboxes are best avoided as they are potentially unreliable and less efficient than a directly driven alternator.

The alternator produces alternating current (AC) of variable frequency depending on the speed of rotation of the wind turbine. Solid state electronic devices called diodes are used to rectify the AC and produce a DC output which is then used to charge a battery.

The power of the wind is proportional to the cube of the windspeed, which means that a doubling of windspeed causes an eightfold increase in energy availability. Therefore strong winds can be a hazard because the high levels of energy they carry can damage the wind turbine. The smallest sizes of wind turbine are designed to survive the strongest winds nature can throw at them, as they can be built strongly enough to spin at very high speed in a storm without damage. This is only satisfactory with rotors smaller than 1m in diameter used in areas not subject to winds exceeding around 50m/s.

Designs for use in windier areas, especially areas prone to hurricanes or typhoons, and all turbines with larger rotors, need some form of overspeed protection mechanism to prevent their destruction in a storm. The most common method is to use a hinged and spring-loaded tail vane which has the effect of turning the rotor wholly or partially edge-

to-the-wind if a strong gust hits the machine.

Siting of a wind turbine

An important point that applies to all wind turbines, is that it is vital to install them where the wind can blow at their rotor across clear and unobstructed ground. Nearby obstructions will produce a turbulent area on their down-wind side which can be extremely damaging to any wind turbine located in such a zone. Therefore wind turbines need to be installed outside the turbulent area on the downwind side of any solid obstruction. This is more to avoid damage from turbulence than because of the obvious loss of energy that also results.

Therefore, wind turbines are best used in open treeless areas, such as deserts, savannah grasslands or on coastlines. This is a simple and basic requirement, but one that is often overlooked.

A minimum mean windspeed in the range 3 to 4m/s is necessary for the effective use of small wind turbines and an annual mean windspeed exceeding 5m/s is preferable for the cost-effective use of larger machines. Meteorological data is commonly available from national meteorological departments and civil aviation authorities, and this can be used to determine the probable mean wind speed in any particular area.

Battery and regulator system

The wind turbine and load need to be adequately sized to keep the battery charged in the least windy periods. Hence a long period of windy weather will tend to cause over-charging of a battery and it is therefore essential to have a charge regulator which senses the battery state-of-charge and dumps surplus power (usually by heating a coil) if the battery is already fully charged. This is more important than with solar photovoltaic systems as the peak output from a wind turbine can be significantly greater than the average output.

It is possible to connect a wind turbine directly to a battery provided that the wind turbine is equipped with an integrated charge regulator to protect the battery from overcharge. Note that in this case the battery is not protected against deep discharge.

If the wind turbine has no integrated regulator, it is possible to install a separate regulator to protect the battery from both overcharge and deep discharge, thus extending its useful life. Although the extra expense of the regulator may not be economic for a small installation (e.g. 50W wind turbine with a 100Ah battery) it will certainly be economically justifiable when it protects a large battery bank.

Installation

Small wind generators for lighting systems are usually quite easy to erect, using a standard 6m length (of 25 to 68mm diameter) water pipe as a mast (see Figure 8.13). If guy wires are used, follow the supplier's instructions. Generally it is prudent to use at least four and preferably five guy wires: if the minimum of three is used, then only one needs to become uprooted to cause catastrophic failure. Proper multistrand guy wires should be used; fencing wire is not good enough as it soon stretches and goes slack. Guy wires are vulnerable to damage from larger livestock, such as cattle, which like to rub themselves against the cables, making them go slack or even uprooting the anchorage. Therefore guyed wind turbines are best fenced off from any livestock.

An alternative is to use a small self-standing tower bolted to a concrete block foundation (perhaps welded up from scrap steel bits and pieces). This can be more robust than relying on guy wires and also it is sometimes difficult to anchor guy wires on rocky ground or very soft peaty ground.

Sizing

The output from a wind turbine, as has already been mentioned, is very sensitive to the mean windspeed. Figure 8.15 indicates the energy per day (kWh/day) that can be expected on average, per square metre of wind turbine rotor. There are two curves as the output is quite variable between different machines and different locations, so the curves are meant to indicate the likely spread between minimum and maximum kWh/day per square metre of rotor.

The equation for calculating the area of a rotor is:

$$A = 3.14 \times r^2$$

where 'r' is the radius (i.e. half the diameter of the rotor). For example, a rotor of 1.1m diameter gives an effective area of 1m².

An example to illustrate how to use the above graph follows. Assuming the yearly mean windspeed is 5m/s, then it can be seen that the mean daily output is likely to be in the range 200-500Wh per day per m² of rotor.

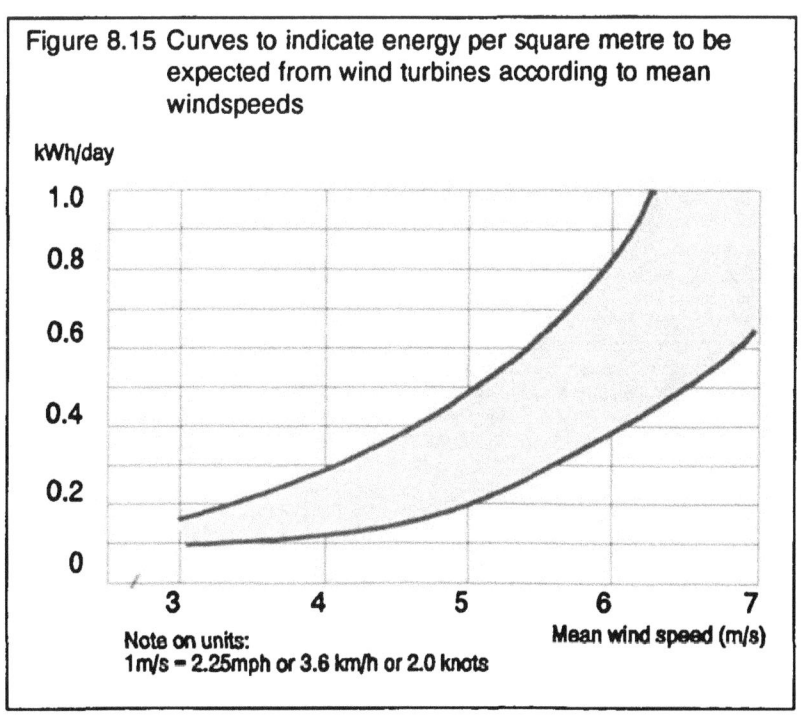

Figure 8.15 Curves to indicate energy per square metre to be expected from wind turbines according to mean windspeeds

kWh/day

Note on units:
1m/s = 2.25mph or 3.6 km/h or 2.0 knots

Mean wind speed (m/s)

As a battery is only about 70% efficient, the useful output from the battery for lighting will be reduced by the battery's internal losses to within the range 140-280Wh/day/m² of rotor. If we assume a safety factor, to allow for seasonal fluctuations, of only 25% of the average energy being available at certain times, then energy in the range 35-70Wh/day/m² of rotor may be confidently expected (and quite a bit more during consistently windy weather). In most cases, three to four days battery capacity is needed to cover calm periods, so the battery will need to have a usable capacity of 100 to 280Wh (8 to 23Ah at 12V). If a lead-acid battery, probably two to three times this capacity will be necessary to ensure it is never over-discharged (unless there is a low-charge regulator).

Around 140 to 280Wh (i.e. 12 to 24Ah at 12V) will comfortably allow 10 to 20 hours per day operation of a 13W tube fluorescent lamp, or almost 5 to 10 hours operation of two such lights or a single 30W fluorescent tube.

If the monthly wind speeds are known, the sizing should be based on the least windy month and using a safety factor of 50% instead of 25%.

Economics

Since a small (50W) wind-charging system (including wind turbine, mast, battery, charge regulator and fittings) can be purchased for approximately US$700 to $1000, this is comparable in cost to a solar photovoltaic system. Obviously, the wind system will perform more effectively and will be more cost effective in windy regions which often lack sufficient sunshine for the use of PV systems.

Selection and Design of Lighting Systems 9

9.1 Introduction

Lighting design is a complex process. No hard and fast rules can be devised which will suit with accuracy all design problems. Furthermore, for precise design work, up-to-date information on light sources, luminaires and control systems should be obtained from manufacturers.

However, the rules that follow are to provide a practical and simple design method for a development worker to design systems for most lighting applications, for example in houses, schools, health centres, hospitals or street lighting.

The following method is primarily intended for electric lighting systems, but to some extent it can also be used for flame-based lighting systems. For this reason, the final section of section 9.3 deals only with flame-based lighting systems.

To avoid doing any calculations or for quick references and checks, use Table 9.4 at the end of this chapter, where full examples of applications and possible solutions are provided.

Check list: Informal definition of lighting needs

At this stage, it is essential that one consider all the following questions:

- What is the precise use for the light?

For example, in a health centre, is it for the delivery room, the surgery or for the rest room? Or if it is for an office, is it for the desk or for general lighting?

- Where is the light needed?

Are the areas that require lighting located in the same building? If not, is it possible to change this by a better organization of the building's use. It may be possible to change the layout of the room so as to make better use of daylight, and by so doing reduce the cost of installation of the lighting system or at least the running cost.

- How often will the light be needed? Once a day? Two hours per day? Once a week? Or only in emergency cases?

- When will the light be needed? During the day? During the night? Both?

- What safety hazards need to be seen clearly? Is a stroboscopic effect likely?

- What are the potentially hostile environmental effects on the lighting system? Dust? Insects? Dripping water? Humidity of the air? Extreme cold or heat?

- What are the energy sources available and when are they available?

- Who is going to use this light, and how?

- Who is going to maintain the lighting system, and how?

- Who is going to install the lighting system, and how?

- Who is going to pay the capital cost of the installation, running costs and maintenance costs ?

- And most importantly, will the users be happy with (and even enthusiastic about) the new lighting system ?

The short and long-term success of the lighting system depends greatly on its acceptance by the user. If a system is at least partly user-owned, user will feel more responsibility for the system and look after it far more effectively.

The questions above do not necessarily need to be answered fully at the outset but need to be kept in mind during the design of the lighting system. These questions can serve as a check list; each question should have a clear answer at the end of the design procedure.

The process of designing a lighting system for a particular need is composed of six stages to be described below:

- Selection criteria;

- Design/sizing;

- Balance of system selection;

- Overall assessment;

- System management;

- Maintenance.

9.2 Selection Criteria

The selection criteria can be divided into four categories. Those related to the:

- Lighting need;

- Energy source;

- Safety aspects;

- Constraints and priorities of both the owner/user and the location.

Criteria related to lighting need

The quantity of illumination in an interior is a function of the nature and duration of the

Figure 9.1 Discussing lighting needs with a local photovoltaic lighting system installer in Lubumbashi, Zaire

visual tasks which are to be performed in it. Other requirements refer to the colour rendering, correlated colour temperature, glare and shadows provoked by the lighting system (see Chapter 2). The 'atmosphere' (or 'ambience') to be created in a working or living interior should also be taken into account as a suitability requirement even in places where it is not considered essential.

Criteria related to energy sources

In Chapters 6, 7 and 8, the sources, storage, cost, practical problems and constraints of energy sources for both flame-based and electric lighting systems have been reviewed. The availability and cost of all energy options should be established for the precise location where the lighting system is to be implemented, as this may strongly influence the process of selection.

Renewable energy sources such as solar and wind energy can offer workable alternatives in many situations. Their potential should be assessed through available meteorological data supplemented as far as wind is concerned by some local measurements. Hydro power and biomass options need careful local evaluations.

As a result of making such an inventory, some lighting options may rapidly be excluded from further analysis.

Criteria related to safety

The presence of a lighting system should not cause risk of injury to its users. It should not present any undue environmental hazards, such as fire or explosion. Finally, the system should allow the users to move about and work effectively without risk of accident, and without causing adverse working conditions.

In the process of selecting and designing, care should be taken to recognize all human and environmental factors which influence safety. Any lighting system which on analysis does not comply with basic safety requirements should be rejected.

Flame-based light sources hardly meet the above conditions, since they have hot parts which may burn if touched accidentally (a potential risk for children). They also present a fire hazard, particularly in thatch-roofed houses which are common in many rural areas. They should not be used in any interior presenting an explosion hazard, for instance a grain milling house where dust can cause an explosion and fire, or a fuel retailing store. Finally, they produce gas emissions which pollute the environment and can be unpleasant and unhealthy.

Constraints and priorities

The objectives and priorities will need to take account of various constraints. Usually the most critical constraints are financial ones. The availability of funds and/or conditions of access to credit must be established.

It is common for the selection process to be limited to a simple comparison of first costs of different systems, but the lighting designer should resist such a simplistic approach and insist on the necessity of taking account of operating and maintenance costs in order to base the choice on the life-cycle cost of each project (see Chapter 10, Lighting Economics).

Energy conservation is also an important consideration that is of increasing concern due to both the cost and the effect on the environment of wasting energy.

Once all the selection criteria have been identified, one can proceed to the design and

sizing of the lighting system. It should be kept in mind that not all the requirements and constraints for a lighting system can be expressed as measurable quantities, in particular those related to the atmosphere, or ambience to be created by the lighting system. This, however, is not a reason to overlook these constraints, because they may be important, or even essential from the users' point of view.

9.3 Design/Sizing

The purpose of the design stage is to:

- Assess the daylight contribution;

- Decide the class of lighting (general, localized, etc.);

- Decide and/or calculate the types and numbers of lamps and luminaires.

Sizing is based on the following sequences of calculations which are described in the flowchart in Figure 9.2. Several methods for determining the numbers of luminaires exist, but the most convenient is the 'lumen method', which aims to achieve a given uniform illuminance on the horizontal working plane.

Electrical Lighting Systems

Stage 9.a Daylight contribution assessment

Daylight contribution should be considered either as a stand-alone lighting system like any other or, more usually, as an additional lighting system when light is needed during the day. That explains why daylight assessment is included in the design/sizing of a lighting system.

In Chapter 5, the advantages and disadvantages of daylighting are detailed. In Appendix C, a method of calculation of the daylight contribution which can be assessed by calculating the Average Daylight Factor (ADF) is given along with an example calculation. However, the evaluation of the daylight contribution can be skimmed most of the time.

Figure 9.2   Design/sizing flowchart

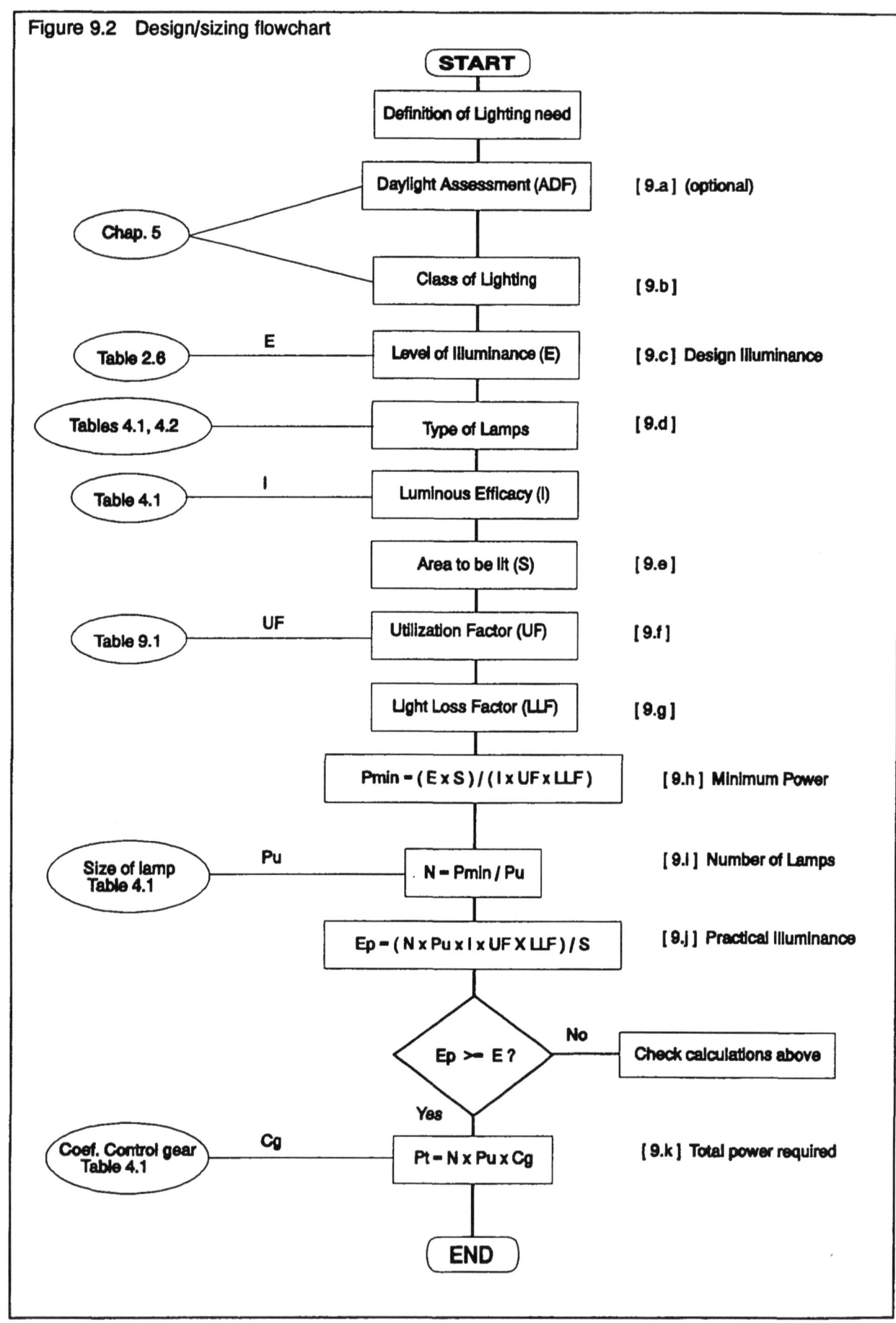

Stage 9.b Class of lighting system

Three different classes of lighting systems can be distinguished in accordance with their layout with respect to the visual task or objects to be illuminated:

- General lighting;

- Localized lighting;

- Local lighting.

The principal characteristics of these classes of lighting systems are described in Chapter 5 'Enhancing Light Sources'. It must be stressed that wherever possible in order to achieve the required illuminance and to save energy it is often better to choose local or localized lighting.

Stage 9.c Level of illuminance (E)

In Chapter 2, Table 2.6 'Choice of design service illuminance' describes illumination levels recommended for various activities and interiors. This table allows one to choose the minimum illuminance, measured in Lux, necessary according to need.

Stage 9.d Choice of lamps

The choice of lamp is made through compiling a list of available and suitable lamps and rejecting those that do not comply with the design objectives. Warm-up time, colour rendering, flicker, life span and luminous efficacy are among the principle criteria which are used to justify the selection or rejection of a lamp.

With the help of Table 4.1 'General characteristics of lamps' and Table 4.2 'Applications of different lamps,' choose the lamps required. Once a suitable lamp is chosen, the luminous efficacy (I), can be determined from Table 4.1 'General characteristics of lamps'. It is essential to choose a lamp which is available on the local market and which has a high luminous efficacy (measured in lumen/W) for a given application. For example, where possible, it is always better to choose fluorescent lamps or halogen incandescent lamps as opposed to standard incandescent lamps.

Stage 9.e Area to be lit (S)

The surface to be lit is often called the working plane (horizontal, vertical or inclined) and is the plane in which the visual task lies (e.g. office desk). For most applications and when no information is available, the working plane may be considered to be horizontal and at 0.7m to 0.85m above the floor.

The bigger the surface to be lit, the larger the lighting system will be, along with increased capital and running costs. Therefore, it is essential to determine the surface where light is needed in the room.

For instance, to provide a general 'ambience' lighting in a living room, the area of the living room itself will be considered in the calculation. If however in the living room, there is a need for a reading place (e.g. a small table with an armchair), the area of this table and immediate surroundings should be determined for the calculation. The area should be measured in square metres (m²).

Stage 9.f Utilization factor (UF)

Only a fraction of the light output of the lamp will be useful on the working plane, either because the luminaires and the walls absorb some of the light output of a lamp or because some light will reach unwanted areas. To take this into account it is useful to estimate the efficiency of a lamp/luminaire combination which can be expressed by the utilization factor.

Definition: The *utilization factor (UF)* is expressed as the ratio of the luminous flux which reaches the working plane to the luminous flux emitted by the lamp.

For instance, a utilization factor will be equal to one for a spot lamp of 20W directly lighting a flat surface situated very close to it, i.e. less than 0.5 metres. Of course, the UF will decrease with distance and eventually be zero if the surface is very far from the light source, i.e. 10 metres.

For each type of luminaire, the Utilization Factor varies with the efficiency, distribution and spacing of the luminaires as well as with the reflectance of the room surfaces and room proportions. It is possible to calculate accurately the UF only if the manufacturer's data for each luminaire are available. In Appendix D 'Method of Calculation of the Utilization Factor,' the accurate calculation procedure for the Utilization Factor is given along with an example. However, in most rural lighting applications, it is sufficient to use the figures in Table 9.1 'Typical utilization factors'. This table gives UF values for the most common cases. For other cases, an extrapolation might be necessary.

Stage 9.g Light loss factor (LLF)

A Light Loss Factor (LLF) should be included in lighting calculations to allow for any degradation of performance with time. Total light losses can be divided into two categories, recoverable and unrecoverable. Losses which can be corrected or avoided by routine lighting maintenance procedures are called recoverable losses. Unrecoverable losses are not subject to correction by lighting maintenance procedures.

The main causes of recoverable losses are:

● Burnouts (applies when a certain number of burned-out lamps are tolerated in an installation before all lamps are replaced);

● Lamp lumen inherent depreciation; (i.e. degradation of the lamp);

● Build-up of dirt on luminaires and lamps which absorbs light;

● Build-up of dirt on room surfaces which affects their reflectance.

Table 9.1 Typical utilization factors assuming that reflectors and diffusers are new and clean

Type of walls and ceiling	Lamp bare	Luminaire with reflector	Luminaire with diffuser*
General lighting such as for classrooms, homes, etc.			
White	0.60	0.70	0.65
Light brown, yellow	0.55	0.65	0.60
Brown, earth colour	0.50	0.60	0.50

Assumptions: the distance between light and working plane is between 1.5 to 2m, the ceiling is not higher than 3m, the room is approximately square.
*The diffuser is backed by a white reflector.

Local, localized and task lighting			
Office desk lamp	N/A	0.80	0.70
Spot lamp	0.80	N/A	N/A

Assumptions: the distance between the light source and working plane is between 0.5 to 1.2m.

Street lighting			
Street Lighting	N/A	0.60	0.50

Assumptions: the distance between the light source and area to be lit is between 2.5 and 3m.

Unrecoverable losses may be caused by:

- Increase of ambient temperature (for fluorescent lamps only);

- Voltage fluctuations;

- Luminaire surface depreciation with age (plastic yellowing, enamel discoloration, etc.).

The LLF can be calculated using methodologies such as those explained in reference documents like the Code for Interior Lighting CIBSE (see Bibliography). However, in most common lighting design calculations, a Light Loss Factor (LLF) of 0.8 is advised.

Stage 9.h Minimum electric power requirement (P_{min})

The minimum electric power, measured in Watts, is determined by inserting the appropriate values of the above factors into the following formula (9.h):

$$Pmin = (E \times S) / (I \times LLF \times UF) \quad (W)$$

where:

    E = average illuminance required on the working plane (lx);

    S = surface area to be lit (m²);

    I = initial bare lamp luminous efficacy (lumen/W);

    LLF = Light loss factor;

    UF = Utilization factor.

Stage 9.i Number of lamps required (N)

According to available sizes of lamps (see Table 4.1 'General characteristics of lamps') and what is available on the local market, calculate the number of lamps required using the following formula (9.i) and round up to the nearest whole number.

$$N = Pmin / Pu$$

where:

    Pu = unit power of lamp (W).

Stage 9.j Practical illuminance check (Ep)

In order to check the actual practical illuminance (Ep) that would be achieved using the chosen lamps, it is advised to use the following formula, 9.j. (Note: this formula is identical to 9.h, but has simply been rearranged.)

$$Ep = (N \times Pu \times I \times UF \times LLF) / S \quad (lx)$$

Where:

    N, Pu, I, UF, LLF and **S** are as above.

Ep should be at least equal or slightly above E (design service illuminance). If not, check all your data and previous calculations. In some cases, rounding up to a whole number (i.e. 1.2 to 2) may lead to unecessary overlighting; in such situations recalculate with another size of lamp to get nearer to the requested design service illuminance (E).

Stage 9.k Total power requirement (Pt)

Some lamps require control gear, some not. (See Table 4.1 'General characteristics of lamps'). Some control gears consume power only while switching on the lamp. Most, however, consume power during the entire time that the lamp is switched on. On average, the control gear consumes between 5 and 35% of the power of the lamp.

For example, a 50W fluorescent lamp can have control gear that consumes up to 8W, therefore the total power consumption will be 58W. For small lamps up to 20W, the percentage may be up to 50%. Unless manufacturers' data are available, it is advised to use a minimum control gear consumption of 25%, i.e. Cg is 1.25. Of course, when there is no control gear Cg is equal to 1. See Table 4.1 'General characteristics of lamps' to see which types of lamp have control gears. Formula 9.k calculates total power needed considering the power of the control gear as well.

$$Pt = Pu \times N \times Cg \quad (W)$$

Where:

    Pt = Total power needed (W);

N = Number of lamps;

Pu = Unit power of each lamp (W);

Cg = Coefficient for control gear.

Luminaire spacing

This section applies only when a uniform illuminance is required all over the reference plane and/or when there is more than one luminaire required. There is a method to calculate the luminaire spacing with accuracy which requires manufacturers' data for each type of luminaire. However in most cases of rural lighting, it is not necessary to calculate these but rather to use common sense. In general, lighting luminaires should not be spaced too far apart or too far from the walls, otherwise a uniform illuminance cannot be achieved.

Comments on the example

In the example on the following page, as light is needed both day and night, it is necessary to calculate the ADF in order to reduce energy consumption during the day. The value found in Appendix C 'Calculation of Average Daylight Factor,' indicates that if there are no buildings or obstacles near the window, the ADF is close to 5%. Hence there is no need for general lighting during the day. For task lighting however, the lighting system will be necessary even during daylight hours.

For general lighting, in order to save energy, a luminaire with a reflector has been chosen (to increase the Utilization Factor). Furthermore two 8W lamps instead of one 20W are chosen. This will allow the use of only one lamp for tasks where less light is needed, for example during labour. Both lamps can then be used for tasks which require more light such as during a delivery.

For task lighting, a technical lamp of 35W with an incandescent halogen bulb with a 38° beam angle will give a concentrated light beam for use during deliveries and also for minor surgical operations. It should be placed on the floor, movable preferably with castors and it should have a movable reflector. The distance between the lamp and the lit area should not exceed 1.2 metres.

Flame-based lighting systems

It should be noted here that it is not usual to calculate flamed-based lighting systems requirements, but instead to buy enough inexpensive kerosene lamps or candles for the required lighting needs. For more expensive systems such as biogas lamps, carbide lamps or kerosene pressure lamps, it can be useful to size the lighting system.

The methodology for sizing a flamed-based system is similar to that of an electric lighting system. However, the calculation is less accurate because the luminous flux of flame-based light is difficult to assess with accuracy. Unfortunately, the manufacturers of such systems do not usually give the lumen output of their products.

Stages 9.a, b, c, e, f and g of the electrical lighting systems methodology are similar to those for flame-based lighting, which allows E, S, UF and LLF to be determined.

Stage 9.l Determination of luminous flux

It is then necessary to determine from Table 3.1 'Performance comparison of various flame-based lamps,' the luminous flux (Iu) of the chosen lamp in order to calculate the number of lamps required and round up to the nearest whole number (Formula 9.l).

$$N = (E \times S) / (Iu \times UF \times LLF)$$

Where:

N = Number of lamps needed;

Iu = Luminous flux of the lamp (lm);

E = Level of illuminance;

S = Area to be lit (m²);

UF = Utilization factor;

LLF = Light loss factor.

Stage 9.m Calculation of illuminance with the chosen lamps

At this stage, it is advised to calculate the level of practical illuminance (Ep) that will be produced by the chosen lamp with the follow-

Table 9.2   Example of electrical lighting design for a health centre delivery room

In a delivery room, there is a need for both general and task lighting as it can be used for routine gynaecological examinations and procedures during the day as well as for deliveries, day or night. Task lighting is necessary for deliveries or minor surgery and should consist of a maximum circular area of 0.6m diameter. The room has the following dimensions: width 3m, height 2.8m, length 3m. The window is: 2m wide x 1.2m high (or two small windows with the same total area) and is made of clear glass. The base of the window is 1.2m from the floor. The walls and ceiling have been freshly painted white. The window faces the front of another building which is 4m high and situated 4m away. This same room is referred to in Appendix C.

Electricity is supplied by a 12 Volt photovoltaic system.

Stage	Comments	Symbol	System 1	System 2
9.a	Calculation of ADF (optional)	ADF (%)	2.4	N/A
9.b	Class of lighting (Chapter 5)		General lighting	Task lighting
9.c	Level of illuminance (Table 2.6)	E (lx)	50	1500
9.d	Type of lamp (Tables 4.1, 4.2)		Fluorescent lamp	Incandescent halogen lamp
	Luminous efficacy (Table 4.1)	I (lm/W)	56	20
9.e	Surface to be lit	S (m2)	3x3 = 9	(0.6x0.6x3.14)/4 = 0.28
9.f	Utilization factor (Table 9.1)	UF	0.7	0.8
9.g	Light loss factor	LLF	0.8	0.8
9.h	Minimum power required	Pmin (W)	(50x9)/(56x0.7x0.8) = 14.35	(1300x0.28)/(20x0.8x0.8) = 28.44
9.i	Size of lamp (Table 4.1)	Pu(W)	8	35
	Number of lamps (Formula 9.i)	N	14.35/8 =1.79--->2	28.44/35 = 0.81 =--->1
9.j	Practical Illuminance (Formula 9.j)	Ep (lx)	(2x8x56x0.7x0.8)/9 = 56	(1x35x20x0.8x0.8)/0.28 =1600
9.k	Coefficient for control gear (Table 4.1)	Cg	1.25	1
	Total power needed (Formula 9.k)	Pt (W)	2x8x1.25 = 20	1x35x1 = 35

Table 9.3   Example of flame-based lighting design: dining and living room

The room is the same as for the previous example on electrical lighting but it is used as a small dining and living room. Light is needed only at night. There is no electricity available but there are biogas or kerosene fuels. Biogas lamps and only ordinary kerosene wick lamps are available.

Stage	Comments		Gas lamp	Kerosene lamp
9.a	Calculation of ADF optional	ADF(%)	2.4	2.4
9.b	Class of lighting Chapter 5		General lighting	General lighting
9.c	Level of illuminance Table 2.6	E(lx)	25	25
9.3	Surface to be lit	A (m2)	3 x 3 = 9	3 x 3 = 9
9.f	Utilization factor Table 9.1	UF	0.7	0.7
9.g	Light loss factor	LLF	0.8	0.8
9.I	Luminous flux Table 3.1	lu (lm)	1000	100
	Number of lamps Formula 9.1	N	(25x9) / (1000x0.7x0.8) = 0.4 --->1	(25x9) / (100x0.7x0.8) = 4.02 --->4
9.m	Practical Illuminance Formula 9.j	Ep (lx)	(1x1000x0.7x0.8) / 9 = 62	(4x100x0.8x0.7) / 9 = 25

Comments on the sizing: In practice, only one or two instead of four kerosene lamps will be used in this room otherwise the quantity of smoke and soot will be too great. Therefore, the illuminance level will be very low indeed. On the other hand, one biogas lamp is sufficient to provide a good lighting ambience.

ing Formula, 9.m (which is identical to Formula 9.j).

Ep = (N x Pu x lu x UF x LLF) / S   (lx)

Where:

   N, Pu, lu, UF, LLF and **S** are as above.

Ep should be at least equal to E (design service illuminance). If not, check all your data and previous calculations.

9.4 Balance-of-System Components

Once the type and size of the lamps have been selected, the choice of the balance of system components should be made (e.g. cables, fittings, etc.).

For any lighting system, the balance of system components (BOS), including spare parts should be selected according to their:

- Reliability;
- Durability;
- Safety;
- Compatibility with any existing installation;
- Availability of spare parts;
- Affordability;
- Possibility of repair and routine maintenance by locally trained people;
- Possibility of local manufacture.

It is beyond the scope of this book to give details on the BOS and spare parts for each type of lighting system. As most of the lighting systems are, or will most likely be, electrical, it is intended here to give some useful hints for the selection of BOS components for them.

Hints for the selection of BOS components

As a general requirement, the current and voltage ratings of cables, fuses and switches must be sufficient to transport safely the current drawn by the lamps including the control gear if any.

Luminaires

The most likely type of system to be chosen will include fluorescent lamp luminaires as these are the most energy-efficient. The luminaire will include the lamp, the ballast/inverter and housing. Further to the general requirements mentioned above, they should embrace the following characteristics:

- Voltage level compatible with the power system;
- Ballast/inverter adapted if necessary to tropical conditions;
- Protection against possible reversing of polarity;
- A minimum luminous efficacy of 50 lm/W;
- Cable connection point should be dust-proof and insect-proof.

Electric cables

It is preferable to use flexible insulated electric cables that are specified to international standards. The diameter of the conductor should be such that the in-line voltage loss is less than 5% at maximum power.

Remember that low-voltage installations (e.g. 12 to 24V) require thicker cables than 220 or 110V AC systems, and that cables should be kept as short as possible to reduce power losses. Finally, it is highly recommended that a local electrical supplier is contacted to ensure that the cables chosen are thick enough for the application.

Junction boxes, switches and wall sockets

These should be chosen such that they can be firmly fixed onto the walls or ceilings. Junction boxes and switches should be weatherproof if mounted outside. There should be at least one switch for each luminaire (for a better energy management). It is not recommended that fluorescent luminaires with an incorporated switch be used because these switches are often of poor quality.

Plugs and wall sockets used to plug movable lamps should be equipped with a system to prevent polarity inversion if the installation is in DC current.

The UK standard 240V (13 A) wall sockets for domestic use are well suited to this usage but at less than 120V (for safety reasons) if DC is used.

Cable-fixing accessories

To prevent electric shocks, accidents or potential vandalism, it is very important that the cables are attached with cable-fixing accessories which should be either:

- Self-locking UV-resistant collars to connect cables together, or onto a mast;
- Self-locking collars with plastic anchors for fixing cables to masonry or earth walls;
- Surface moulded wiring cable clips with hardened steel nails for fixing cables to wooden walls.

9.5 Overall Assessment of the Design

After the completion of the design of an installation, detailed calculations are required to determine costs and energy consumption in order to verify how they comply with the initial economic and energy budgets.

Financial evaluation

Taking into account that in absolute terms it is difficult to judge the economics of an installation, one should therefore compare the proposed design with the existing one (if any), or with alternative technologies. For the comparison to make sense, all alternatives should provide the same lighting levels.

Life-cycle costing (see Chapter 10) provides the best method for comparing technically equivalent alternatives. It allows the judgement to be based not only on the capital costs of the installation, but also on the energy costs and the operation and maintenance costs which will be involved during the whole useful life of the competing systems.

Energy consumption

Any new lighting design should not waste energy: indeed minimizing the energy consumption should be a primary objective. However, the cost of the energy consumed will probably be the factor of most concern to the user. It makes sense not only to use energy-efficient lamps, but also to make the best use of daylighting. This is particularly important for lighting systems powered by renewable energy sources, since it may be possible to reduce the initial capital investment.

When the grid is available, any new design should seek to take advantage of all the benefits of the electricity tariff system, or at least avoid its disadvantages.

System effectiveness and acceptability

After the system is installed, an assessment of its effectiveness and acceptability should be undertaken. If a luxmeter is available, effec-tiveness can be assessed through a photometric survey of the lighting levels actually achieved by the installation. Results of this survey are then compared with the initial specifications in order to establish the extent to which the installation meets these initial specifications.

Acceptability is assessed by means of discussions with the users, in order to know to what extent their requirements and expectations are met by the installation. It should be noted that electric lighting systems newly installed in rural regions are likely to be the first such systems in the area, and so they have every chance of being well accepted!

9.6 System Management

Sound system management is the key to sustained performance of any lighting system. Its aim is to ensure the efficiency of system operation at all times.

This objective is attained by means of a good system control to help with energy conservation. The control can be manual or automatic with large and complicated systems.

In all cases, a good training programme is required for staff or family members to ensure that they have an awareness about energy conservation. This can be reinforced by visual aids (posters, labels, and so on) placed in appropriate locations. Posters and labels

---

Example of content for a poster appropriate for a photovoltaic lighting system

♦ Weekly cleaning of array

♦ Cutting trees or branches that cause shading of the array

♦ Turn off any lamp not in use

♦ Clean the luminaires frequently

♦ How and when to top up the electrolyte level in batteries

♦ Never add acid but always distilled water to the batteries

♦ How to contact the technician on failure of system.

---

must be attractive and if possible in the local language. They must be washable, made of durable materials so that they last at least as long as the lighting system and should be fixed on walls or lighting components.

Manual control is performed by the users of the system. It is cheap, but requires a great deal of switching flexibility in order to be efficient. Automatic control can be achieved by time switches turning the light off after working hours, by photocells monitoring daylight level in order to switch off or dim part or all of the luminaires, and by occupancy detectors controlling the light according to the presence or absence of occupants. Automatic control should always allow for manual overriding in case of technical or other unexpected problems.

9.7 Maintenance

Time, dust and insects are the greatest enemies of a lighting system. The latter two can degrade the light output of an installation in a matter of weeks, e.g. when insects make their nests inside a luminaire.

It should be stressed that regular maintenance of the lighting system is necessary in order to ensure the continuing quality of its performance and the safety of its operation. Proper maintenance entails:

Figure 9.3 Routine maintenance of two lighting systems, Shaba, Zaire

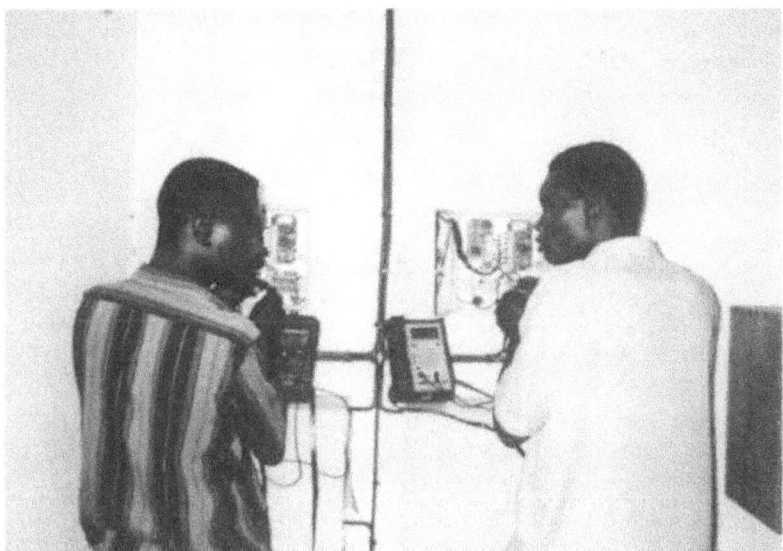

- Relamping (i.e. replacing old or failed bulbs). Always replace a failed lamp by one of the same power and, whenever possible, of the same brand name or with a compatible one. This is particularly important for discharge lamps;

- Replacing faulty or unsafe components;

- Cleaning lamp and luminaire (when off and cold, with soapy moist sponge and then with a dry cloth);

- Cleaning or painting the room surfaces with white paint especially when light reflected from the room surfaces makes an important contribution to the interior lighting (e.g. general lighting);

- Checking all BOS components of the lighting systems e.g. switches, cables, fuses, pipes, etc.;

- Checking energy consumption of the system;

- Keeping a log-book of any tasks performed.

For small lighting systems, like a private house, shop or a small health centre, servicing can be (and usually is) done on an *ad hoc* basis, which means that a lamp is replaced when it burns out, and, similarly, cleaning is done only when necessary.

For important lighting installations, alternative servicing or 'planned maintenance' schedules should be developed in order to establish the most economic maintenance programme.

The maintenance requires labour, tools and spare parts. Maintenance training during the installation of lighting systems, and on a regular basis afterwards, is necessary to ensure that the persons in charge of the maintenance are able to perform it properly. A set of tools including screwdrivers, brushs, spanners and, for large installations, an electrical tester or multimeter should be made available. Finally, spare parts need to be available at any time.

9.8 Examples of Electrical Lighting Solutions

Table 9.4 Examples of electrical lighting solutions

Type of location	S area to light	D*	E Minimum pre-elect-trification	Example of solutions with minimum power rating lamp
	m2	m	lx	
EDUCATION / BUSINESS				
Office desk	2	0.7	100	1 x 8W fluorescent tube with reflector
Office ambience	10	1.8	15	1 x 6W fluorescent tube
Reading area	1	1.8	50	1 x 4W fluorescent tube
Classroom	40	1.8	50	7 x 11W fluorescent tubes
Workshop ambience	20	1.8	20	1 x 8W fluorescent tube
Work spotlight	1	1	20	1 x 6W fluorescent tube or 1 x 15W Incandescent halogen with reflectors
HEALTH CARE				
Examination room spotlight	2	1	500	1 x 25W fluorescent tube with built-in reflector on stand
Delivery room spotlight	0.2	1	1500	1 x 35W 12V halogen with built in reflector on stand
Delivery ambience	10	2	50	1 x 20W fluorescent tube
Post delivery room	20	2	15	1 x 11W fluorescent tube
Surgery spotlight	0.2	1	1500	1 x 35W 12V halogen with built in reflector on stand
Surgery ambience	10	2	50	1 x 20W fluorescent tube
Rest room	20	2	15	1 x 11W fluorescent tube
DOMESTIC LIGHTING				
Dining area	12	1.8	25	2 x 8W fluorescent tubes
Living area	12	1.8	15	1 x 8W fluorescent tube
Bedroom	10	1.8	15	1 x 8W fluorescent tube
Kitchen area	8	1	25	1 x 8W fluorescent tube
Bathroom sink	2	1	100	1 x 8W fluorescent tube
Toilet	2	2	10	1 x 3W incandescent bulb
OUTDOOR LIGHTING				
Street / outdoor	250	3	2.5	1 x 18W low-pressure sodium luminaire
Sign illumination	0.5	0.2	200	1 x 8W fluorescent tube

*D = Distance between light source and surface to be lit.

Assumptions
Walls and ceilings are freshly painted in white.

Lighting Economics 10

10.1 Introduction

The major concern of most groups or individuals who may want to improve their existing system of lighting will be 'how much will it cost?' This chapter looks at the economic factors involved in choosing between different lighting systems and power sources, and examines some of the ways in which rural communities can fund such improvements. Every situation and every lighting application is different, and so it is impossible to say which particular type of system will be more economic in all cases: however, simple methodologies can be defined that can be easily applied to the user's situation, whether it be the choice between two types of portable lamp or the cost of lighting a whole hospital.

The uses and limitations of economics

Consider a case where the user wants to buy a new kerosene pressure lamp, and the choice is between a cheap one that uses a certain number of litres of kerosene per week, and a more expensive one that provides the same light but uses slightly less fuel. Economics provides the framework to determine which will cost less in the long run.

Most economic analyses take both costs and benefits into account. For lighting this presents a problem because the benefits are often qualitative and cannot be expressed in monetary terms. Even where there are economic benefits, they are usually indirect and difficult to quantify.

In workshops, stores, offices and bars, better lighting results in increased productivity and sales. In health centres and schools, it allows better efficiency and improved quality of services. In households, good modern lighting means better safety, a generally improved quality of life, and more time for homework, handicrafts or hobbies.

Solar lighting at Tourakoro, Mali

The installation of just three 13W, fluorescent electric lamps powered by a photovoltaic module in a classroom at Tourakoro, a remote Malian village, caused a minor revolution in village life:

+ teachers could provide evening courses to pupils preparing for national exams, and as a result the latter performed for the first time with 100% success.

+ they could also initiate a literacy course for adults who were too busy to attend daytime courses. The class became the common place for all the evening meetings and social events.

+ at the approach of the Ramadan, when every villager rushes to his tailor to order the traditionally required new clothes, the lit class room offered a night-time working place to the tailors, who could then satisfy all the orders on time: previously, many people could only get their new clothes on their way to the mosque in the Ramadan morning, and many others could only get theirs after the festival.

A study of the effect of rural electrification for small businesses in Bangladesh found that 98% of those surveyed believed that electricity had increased their income. These were mainly small shops selling food, household goods or services, and most used electricity to power two or three lamps.

It is possible to make a decision between two systems based on purely economic grounds only if the quality of lighting service from those two systems is identical. In reality, this will rarely be the case, and when a comparison is being made between flame-based and electrical lighting it is very far from true. The user will need to consider the economics along with other factors to decide whether the investment in a new system is worthwhile.

When carrying out economic calculations it is always better to use local prices and costs. However, these may not always be available so this book gives international prices in US dollars for most systems and components in the relevant chapters. For quick reference, Table 10.6 at the end of this chapter gives approximate prices for the various different types of lamps and bulbs.

10.2 Payback Period

One of the simplest ways of making an economic assessment of various options is the payback period method. This is of use in cases where changing from one system to another would involve a certain investment, but would also bring certain savings. It is a measure of how long it takes the benefits to pay back the investment, so the greater the benefits the shorter the payback period. This is best illustrated by the following example which involves changing a standard incandescent bulb for a compact fluorescent lamp.

Example: CFL vs Incandescent bulbs

Compact Fluorescent Lamps (CFLs) represent a great improvement over standard incandescent lamps (SILs) both in terms of efficiency and durability. Unfortunately CFLs are presently quite expensive and many people ask themselves whether it is worthwhile investing in them.

In this example the payback period will be determined for a situation where a SIL of 75W

Table 10.1 Payback example: incandescent vs CFL bulbs		
Price of electricity	0.10 $ / kWh	
Hours of use	4 h / day	
Annual use	1460 h / year	
Annual electricity cost	Incandescent	CFL
Circuit watts	75 W	15 W
Electricity used per day	0.30 kWh / day	0.06 kWh / day
Electricity used per year	109.5 kWh / year	21.9 kWh / year
Electricity cost	10.95 $ / year	2.19 $ / year
Average annual bulb cost		
Life of incandescent bulb	1000 hours	8000 hours
Average number of bulbs per year	1.5 per year	
Cost of bulbs	0.50 $	
Average cost per year	0.75 $ / year	
Total running cost	11.70 $ / year	2.19 $ / year
Annual saving	**9.51 $ / year**	
Investment (CFL) cost	20.00 $	
Payback period	**2.1 years**	
Life of CFL (at 4hr / day)	5.5 years	

is changed for a CFL of equivalent light output (i.e. 15W). The system is assumed to run from the grid, and the lamp in question is used for four hours per day. The calculation is shown in Table 10.1.

The investment cost is the price of a 15W CFL, taken as US$20. Each year there will be certain savings made due to the lower electricity consumption of the CFL, and so it is necessary to calculate the annual running costs of both the SIL and the CFL and find the difference between them.

The incandescent lamp, at 75W and four hours per day, consumes 109.5kWh per year. At US$0.10/kWh, which is by no means expensive for grid electricity, this equates to an electricity cost of US$10.95 per year. In addition, the bulb will need replacing about every 1000 hours. As four hours of use per day amounts to 1460 hours per year, this means that on average about 1.5 bulbs per year will be needed (i.e. two every three years). At US$0.50 per bulb this adds US$0.75 to the

annual running costs, bringing them to a total of US$11.70 per year.

The CFL also runs for four hours per day, but because it consumes only 15W the annual electricity cost (at US$0.10/kWh) is only US$2.19 per year. Therefore there is an annual saving of 11.70 to 2.19 - US$9.51 per year. If the investment cost (US$20) is divided by the annual saving (US$9.51) this gives a payback period of just 2.1 years. The lifetime of the CFL is about 8000 hours, and so at 1460 hours of use per year it can be expected to last 8000/1460 - 5.5 years. As long as this is longer than the payback period (which in this case it obviously is) then the original investment is worthwhile.

Sensitivity analysis

The above example has been extended to give the payback period for a range of unit electricity costs and for various numbers of hours of use per day. Figure 10.1 shows the results of this analysis, and clearly illustrates

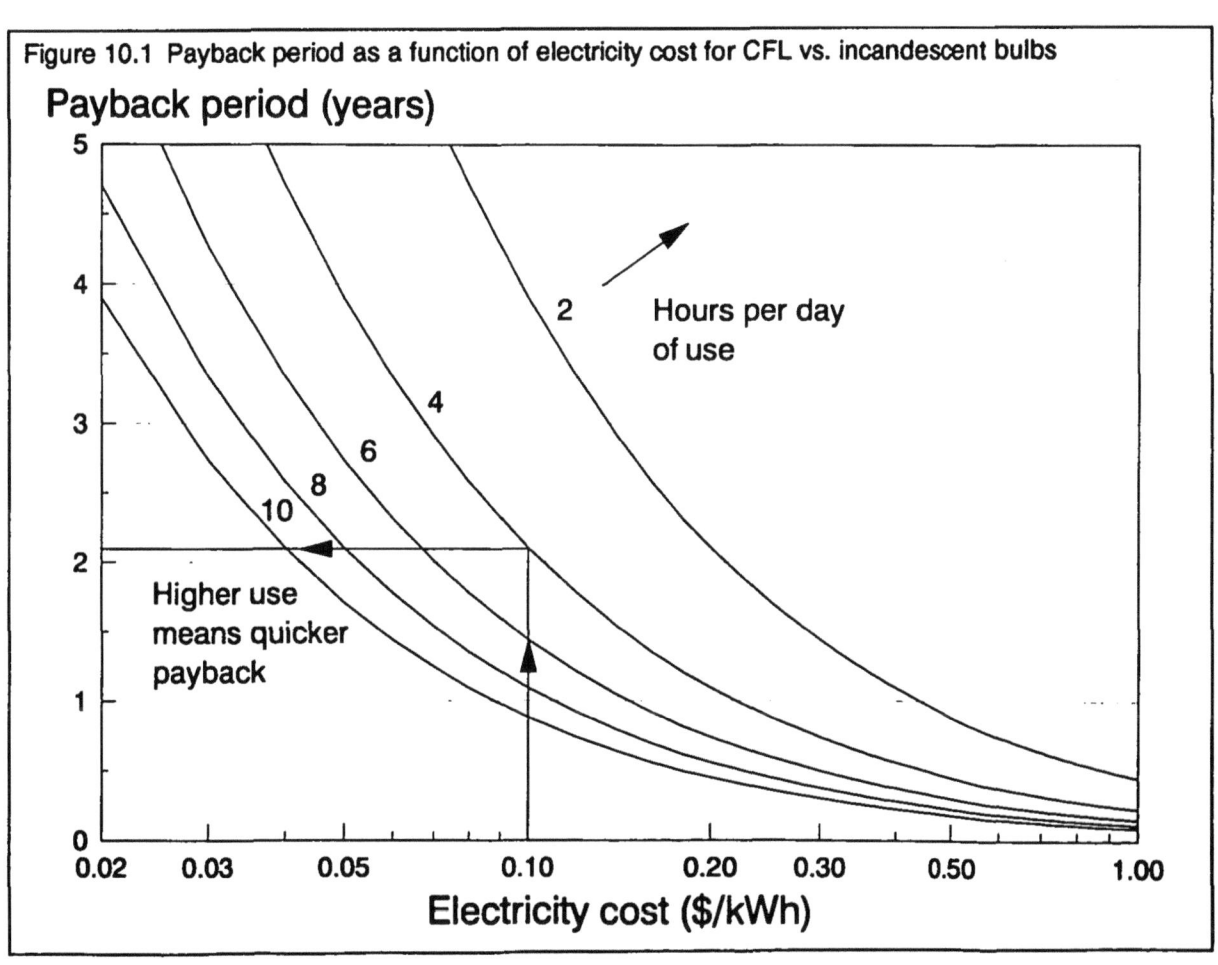

Figure 10.1 Payback period as a function of electricity cost for CFL vs. incandescent bulbs

that the payback period is shorter for higher electricity costs and longer hours of daily use. This is because the bulk of the savings are due to the lower energy consumption of the CFL, and so the greater the use, the more is saved (compared to the incandescent bulb) and the faster the cost of the CFL is paid back.

At higher numbers of hours use per day, the calendar life of the CFL becomes shorter and, as described above, may not exceed the payback period. For instance, 10 hours use per day corresponds to 3650 hours per year. Therefore the CFL calendar lifetime is 8000 / 3650 = 2.2 years. Comparing this with the '10 hours per day' line on the graph, this means that the CFL would only be a good investment in this case when electricity costs are higher than about US$0.04/kWh.

It may be noted from the graph that the unit electricity costs are shown up to US$1.00 per kWh. Although this is far above the cost of grid electricity in most countries, it is not unrealistic for the unit power cost from a small diesel generator or other remote power source.

In summary, payback period is a quick and easy method that can be used in simple situations where one system is being substituted for an existing one to give a saving in running costs. For more complex situations, and where costing is required in terms of cost per year over long periods, the more correct method is to use life-cycle cost analysis.

10.3 Life-Cycle Costing Methodology

The conventional method for comparing technically equivalent alternatives is based on life-cycle costing (LCC). It can be used to compare two or more systems, or just to work out the costs of a single system. The LCC is calculated for each system being considered and is the sum of all the costs associated with the initial purchase, installation, operation and maintenance of a system throughout its operational life. The period normally considered for such a comparison is the lifetime of the longest lived system or component being compared. Life-cycle costing can be used to compare any lighting systems for

which costs and lifetimes are known, and can be reduced to an effective cost per year for each system.

The resulting life-cycle cost is much more informative for decision-making than simply considering initial costs. Generally, equipment using 'consumables' such as fuel or primary cells tends to have low first costs and high running costs, while equipment using renewable energy resources tends to appear expensive in terms of first costs but can in fact, over the years, be more economic through having low or even negligible running costs.

Another advantage of life-cycle costing is that it takes account of the longevity of the equipment. For instance in a comparison between a kerosene lamp and a solar PV lamp, the PV lamp may last 10 years, whereas the kerosene lamp may last only two years. Therefore the analysis period is taken as 10 years, and the cost of five replacement kerosene lamps are included over this period.

The life-cycle cost of a projected equipment procurement is the sum of the following components over the whole of the analysis period.

Life-cycle cost = **capital cost**
+ **replacement costs**
+ **annual cost**
+ **residual costs**

where:

- The **capital cost** is the initial outlay to establish the system including purchase, shipping, transport and installation of the equipment.

- **Replacement costs** cover any components that it is estimated will need replacing within the period of the analysis. For kerosene lighting the whole lamp will need replacing every two or three years. For electrical systems, replacements might include fluorescent tubes or batteries.

- **Annual costs** cover any operation, routine maintenance and fuel costs. This might include fuel for kerosene and gas lamps, dry cell batteries for torches, candles and costs for battery charging from a central charging station. Fuel costs are not relevant to renewable energy sources. It could also include annual expenditure on items that are replaced regularly such as wicks and glasses for kerosene lamps. Items replaced at intervals of one year or less are also best dealt with under annual costs. Annual costs should be summed over the whole analysis period.

● **Residual costs** are the costs of disposal of the equipment (e.g. recycling lead), and /or the income from any resale or scrap value that it may have at the end of the analysis period. For small lighting systems this is usually negligible and, for simplicity, has been assumed to be zero in all of the following examples.

Table 10.2 shows which of these costs can apply to different systems.

A life-cycle cost analysis is straightforward to perform, but it is useful to approach it in an organized manner to avoid confusion. Collect together all the data that is needed before beginning:

● Cost of initial system(s);

● Cost and estimated lifetime of components;

● All the annually recurring costs.

Where possible use local data and costs. However, if this is not available, the various chapters in this book give indicative figures for most costs. When estimating lifetimes it is only possible to be approximate. The lifetime should reflect the quality and hence the price of the product (e.g. a cheaper kerosene lamp may last two years, but a more expensive model may last four years). The analysis period will usually be the lifetime of the longest lived component. In some cases it may be simpler to choose a period that is a multiple of the estimated component life-times, as long as it is realistic. For instance, if the lifetime of a kerosene lamp is estimated at three years, and the longest lived component (say a small PV module) has a life of about 10 years, then it may be more sensible to choose an analysis period (and module lifetime) of 12 years, so that lamp replacements can be placed evenly at years 3, 6 and 9.

Discounting of future costs

When adding future costs (i.e. fuel, operation and maintenance, replacements) into a life-cycle cost analysis, special care must be taken. To make a meaningful comparison all future costs should be 'discounted' to their equivalent value in today's economy, called their 'present worth' or PW. To do this each future cost is multiplied by a 'discount factor' calculated from the discount rate. A rate of around 10% may often be valid for approximate calculations in 'hard currencies'.

Table 10.2 Components of the life-cycle cost for various lighting options			
	Capital costs	Replacement costs	Annual costs
Candles	None	None	Cost of candles for a year
Kerosene lamp	Lamp	Replacement lamps	Fuel, wicks & glass mantles
PV lighting	PV kit or portable lamp	Battery, tubes & regulator	Negligible (distilled water for battery)
Battery charging at recharging station	Battery	Battery Bulbs	Charging and transport costs

Money available today is worth more than money in the future and the discount rate is the rate at which the real value of money would increase if invested. A discount rate of 10% means that in real terms it makes no difference to a user whether he has $10 now or $11 dollars in one year's time. Therefore a cost of $11 in one year's time has a present worth of $10.

The concept of discount rates can be confusing but the central point is that the further in the future the cost, the lower its present worth and the less impact it has on the total life-cycle cost. Discount rates are usually taken as between 8 and 12% per year for hard currencies.

Calculation of present worth

There are two types of calculation that are useful to express a future cost as its present worth.

- The first is used to calculate the present worth of a *single payment* at a certain date in the future, say the replacement of a battery after five years.

- The second is used to calculate the total net present worth of a *recurring cost*, such as annual fuel or maintenance costs. This is effectively the sum of many discounted single payments over the analysis period.

To avoid the need for complex equations, the relevant PW can be found by multiplying the actual cost by a factor that can be found in the tables in Appendix E. It is not normally necessary to consider inflation, as it is assumed that costs and benefits move with the general inflation rate, and so their cost in real terms stays the same.

Examples of both calculations are given in the boxes on the following page.

Annualized life-cycle cost (ALCC)

This is the total LCC expressed in terms of a cost per year. This is far more useful that the raw life-cycle cost, as it gives an indication of what the user will actually need to afford if

the system is chosen, and financing can be obtained.

However, the LCC cannot simply be divided by the number of years in the analysis as this takes no account of the change in value of money due to discounting. The LCC must instead be divided by the factor Pa from Table E.2, found using the chosen discount rate and a number of years corresponding to the analysis period. The annualized cost is expressed in $/year for each system.

Other measures of worth

To try and make a more meaningful comparison between systems that provide different levels of service, some other measures of cost effectiveness can also be used.

The life-cycle costs can be expressed as a cost per hour of illumination. This is simply done by calculating the estimated total number of hours of use per year, and dividing the annualized LCC by this number to give a result in cost per hour. It is often found that when users upgrade to a system which gives an improved lighting service, they use it for more hours per day. The cost per hour measure allows different systems to be compared in this case.

It is also possible to take the quality of the light into account by expressing the ALCC as a cost per lumen per unit time, or kilolumen-hour. To do this, multiply the luminance (in lumens) by the hours of service per year, and divide the ALCC by the result. For instance, a kerosene pressure lamp with a 400 lumen output might be used for three hours per day, which equates to 1096 hours per year. Therefore the total light output over the year is 1096h x 400 lm - 438,000 lm-hour or 438klm-hour. If the ALCC is US$80, the the cost per kilolumen-hour is US$80/438klm-hour = US$0.18/klm-hour.

The latter method emphasizes the efficiency and superiority of service of electrical lighting (particularly using renewable power sources) over flame-based sources. However, it is of limited practical use, as a community with limited disposable income will be more inclined to consider the absolute cost per year, or the cost per hour for a given requirement.

Present Worth Calculations

(a) Single payment

For a single future cost **Cr**, payable in **N** years of time, the Present Worth **PW** is given by:

$$PW = Cr \times Pr$$

Example : It is estimated that a new 100Ah solar battery will be required for a lighting system in five years time. It is assumed that a new battery presently costs US$80 and that battery prices do not change relative to general inflation, and that the discount rate is 10%. Using Table E.1, in Appendix E, with **d** = 0.10 and **N** = 5 (the number of years hence that the payment is to be made), this gives a discount factor **Pr** of 0.62. Therefore the present worth of this future cost is:

$$PW = \$80 \times 0.62 = \$50$$

(b) Annual payment

For a payment **Ca** occurring annually for a period of **N** years the net Present Worth **PW** is:

$$PW = Ca \times Pa$$

Example : The kerosene cost for a particular household is US$2 per week, which makes US$104 per year. Assuming a discount rate of 10% and a length of analysis of 10 years, Table E.2, in Appendix E, (with **d** = 0.10, **N** = 10) gives a cumulative discount factor **Pa** of 6.14. The net present worth of the kerosene costs is therefore:

$$PW = \$104 \times 6.14 = \$639$$

In electrical lighting situations where the choice is only between different power sources, and the actual luminaire configuration is the same, it makes more sense to do the analysis in terms of cost per kWh of energy. To do this, calculate the annualized life-cycle cost for just the generation system, and divide this by the yearly power demand of the lighting load in kWh.

10.4 Examples and Case Studies

The method described above will be demonstrated in the examples that follow. The data are taken from real case studies, and are intended to represent typical situations. In some of the cases the analysis has been extended to look at a range of conditions.

Household lighting

As a worked example, a comparison has been made for a household lighting situation. The lighting demand is four hours per night and the comparison is between either using two kerosene hurricane lanterns or using a small solar PV lighting kit having two 8W fluorescent tubes.

First, it is useful to collect together the relevant price and lifetime data as shown in Table 10.3. An analysis period of 12 years has been chosen as this approximately corresponds to the likely useful lifetime of the PV module, the longest lived component, and is easily divisible by the battery life and lamp life, etc. A discount rate of 10% (or 0.10) has been chosen, as this is usually considered typical for developing countries (if working in hard currency).

The data in Table 10.3 can then be used in the life-cycle costing, as shown in Table 10.4. For clarity this has been laid out in four sections, dealing with capital cost, replacement costs, annual costs and result. The method is explained below.

Capital costs

For the flame-based lighting case the capital cost is simply the cost of the two hurricane lamps at US$8 each. For the lighting kit, it is

Table 10.3 Data for domestic lighting life-cycle cost comparison		
General data		
Analysis period	12	years
Discount rate	10	%
Lighting demand	4	h / day
Hurricane lamp data (per lamp)		
Capital cost	8	US$
Life	3	years
Fuel consumption	0.03	l / hour
Fuel price	0.8	US$ / l
Light output	50	lm
PV lighting kit data		
Capital cost	350	US$
Light output (per lamp)	400	lm
The kit is comprised of		
Item	Replacement cost	Life (years)
PV module 18Wp	0	12
Battery 70Ah	70	3
Tubes (2 x 8W)	8	3
Regulator	100	6

the cost of the total kit (i.e. photovoltaic module, battery, charge regulator, two lamps) including components for installation (cables and fittings, etc.). The specifications quoted are from a commercially available kit, whose capital cost is about US$350. In this case it is assumed that the user would do the installation himself, and therefore no direct costs are incurred. For a larger example (say an electrical lighting system for a whole hospital) installation costs must be taken into account in the capital costs section.

Replacements

For the flame-based lighting case, the lifetime of the hurricane lamps is about three years, and so within the 12 year analysis period they will be replaced at years 3, 6 and 9. Because these costs are in the future, they should have the appropriate discount factors applied

Table 10.4 Life-cycle cost comparison for domestic lighting case

Flame-based household lighting	PV lighting kit

Capital cost

Hurricane lamp x 2	16.00 $

Replacement costs

Both lamps   16.00 $

at year	factor Pr	PW
3	0.75	12.00 $
6	0.56	8.96 $
9	0.42	6.72 $

Net PW of replacements   27.68 $

Annual costs

Fuel

Fuel use (both lamps)	0.06	litres /hour
Hours per day	4.00	hours
Weekly consumption	1.68	litres / week
Kerosene costs	0.80	$ / litre
Weekly fuel bill	1.34	$
Annual fuel bill	69.89	$

Non-fuel running costs
Glasses & wicks	20.00	$ / year

Total annual costs	89.89	$
Factor Pa over 12 years	6.81	
Net PW of annual costs	612.47	$

Results

Total life-cycle cost	656.15	$
Factor Pa over 12 years	6.81	
Annualized cost	96.30	$ / year

Cost per hour of lighting	0.07	$ / hour
Light output	100	lumen

Cost per light output	0.66	$/klm-hour

Capital cost

Complete kit with 2 lamps	350.00 $

Replacement costs

Battery at   70.00 $

at year	factor Pr	PW
3	0.75	52.50 $
6	0.56	39.20 $
9	0.42	29.40 $

Tubes x 2 at   16.00 $

at year	factor	PW
3	0.75	12.00 $
6	0.56	8.96 $
9	0.42	6.72 $

Regulator at   100.00 $

at year	factor	PW
6	0.56	56.00 $

Net PW of replacements   204.78 $

Annual costs

Fuel	0.00 $

Non-fuel running costs	0.00 $

Total annual costs	0.00 $
Net PW of annual costs	0.00 $

Results

Total life-cycle cost	554.78	$
Factor Pa over 12 years	6.81	
Annualized cost	81.42	$ / year

Cost per hour of lighting	0.06	$ / hour
Light output	800	lumen

Cost per light output	0.07	$/klm-hour

(1 klm = 1000 lm)

to them so that they are expressed in their present worth. Taking the first replacement at year three, the discount factor is Pr, for a single future payment, and can be found in Table E.1 in appendix E using N = 3 and a discount factor, d = 0.10, which gives a factor Pr = 0.75. The present worth (PW) of this replacement is therefore the cost of the two hurricane lamps multiplied by the discount factor (0.75 x US$16 = US$12). This process is repeated for years 6 and 9, and then all three replacement costs are summed to give a net present worth of US$27.84.

For the PV lighting kit the process is the same. Both the battery and fluorescent tubes have about a three year life and so must be replaced at years 3, 6 and 9, using the appropriate discount factors. The regulator is likely to need replacing during the analysis period and so has been replaced at year 6.

Annual costs

For the flame-based case it is first necessary to calculate the fuel costs. Knowing the consumption rate (0.03 l/h x 2 lamps = 0.06 l/h in total) the consumption is 0.06 l/h x 4h x 7days = 1.68 l/week. With a kerosene cost of US$0.80/l, the yearly fuel bill is US$0.80/l x 1.68 l/week x 52 = US$69.89.

There will also be regular running costs every year for small replacement items such as glass mantles and wicks, estimated at US$10/year x 2 lamps = US$20 per year. Therefore the annual bill for both fuel and other running costs is US$89.89.

To work out what contribution these future annual costs will make over the entire 12 year analysis period the *cumulative* discount factor Pa must be used. This is found in Table E.2 in Appendix E, using N = 12 years, and discount rate d = 0.10, to give a factor, Pa = 6.81. Hence the net present worth of the annual costs is found by multiplying the annual costs by the cumulative discount factor, US$89.89 x 6.81 = US$612.47. The PV lighting kit has no fuel costs, and no running costs.

A larger system for community use (for both electrical and flame-based lighting) that is part of a planned maintenance programme should have annual maintenance costs included in the annual costs.

Results

The first item to be calculated is the total life-cycle cost for each system, which is the sum of the capital, replacement and annual costs. To find the annualized cost (ALCC), the total life-cycle cost must be divided by the cumulative discount factor Pa, with N = 12 years and discount rate d = 0.10. As found above (from Table E.2) this is 6.81.

The ALCC for the kerosene and PV cases are US$96.30 and US$81.42 per year. These numbers represent the sum that the householder must be prepared to pay per year for the purchase and running of each system. Although PV comes out slightly cheaper, the small difference between the costs in this case is not really significant as this type of life-cycle costing estimate is by no means accurate. It would therefore be said that the annualized costs are comparable in this case.

To take the analysis a step further the cost per hour is calculated. This is simply the annualized cost divided by the number of hours of lighting during the year (4h/day x 365 days). As the annualized costs are very close, so the costs per hour of US$0.07 and US$0.06/h (for the kerosene and PV systems respectively) are also very similar.

The last line gives the cost per unit of light output, measured in dollars per thousand lumen-hours or US$/klm-hour. The cost per hour of lighting is divided by the light output in lumen. The PV system has a cost less than one eighth of that of the kerosene lamp system when measured this way. This is because its lumen output (400 x 2 = 800 lm) is so much better than that of the two kerosene lamps (50 lm x 2 lamps = 100 lm).

Sensitivity analysis

The above example is valid for one particular set of circumstances and prices and is fairly straightforward to work out by hand for a given situation. In this case the annual cost of the PV system turned out very similar to that of the flame-based system, (however the PV system gives eight times more light!). Because most of the LCC of the PV kit comes from the capital and equipment replacement costs, the final annualized cost is fairly well defined. However, for the flame-based case

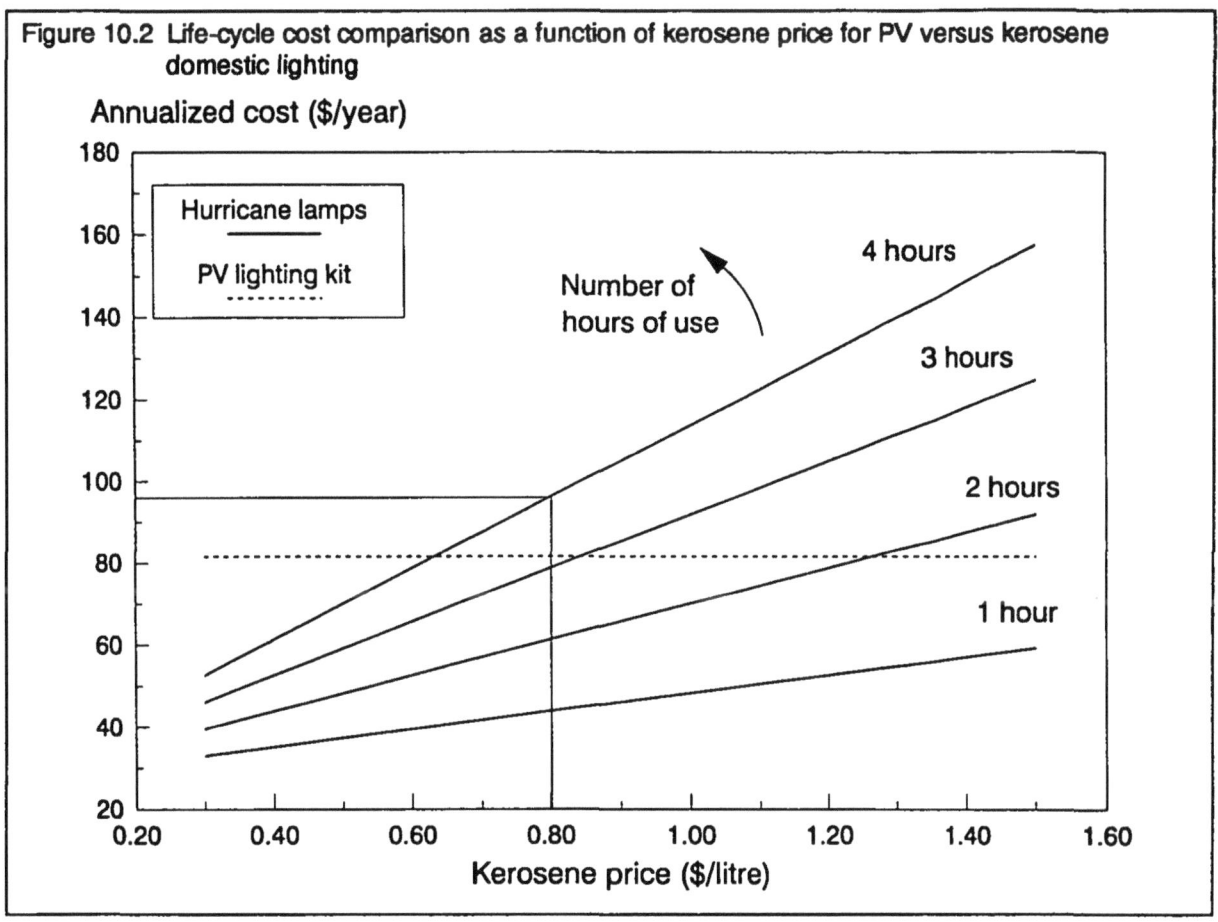

Figure 10.2 Life-cycle cost comparison as a function of kerosene price for PV versus kerosene domestic lighting

the bulk of the cost is for fuel, and so the number of hours of use and fuel costs will have a large impact on the annualized cost.

This is illustrated in Figure 10.2, which shows how the annualized cost of the flame-based system varies with kerosene price. The different sloping lines represent different numbers of hours of lighting, and the straight broken line represents the PV kit. This makes it easy to see whether, in a particular situation, the cost of kerosene lighting is greater or less than the cost of PV lighting. (The annualized PV kit cost has been assumed to be constant for the sake of simplicity, although in reality there would be some slight variation due to more frequent tube and battery replacements with longer hours of use).

It is interesting to note that in this example there are very few cases (except at very low kerosene price or low daily use) where kerosene lighting would be cheaper than using the PV kit. In general, the greater the lighting demand, the more expensive flame-based lighting becomes.

Portable lamps and lanterns

The general method of life-cycle costing has been applied to various types of lanterns and lamps popular in rural areas, based on four hours of lighting per day. Table 10.5 shows the capital cost, annualized life-cycle cost and cost per klm-hour of candles, kerosene lamps, butane lamp (with rechargeable cylinders) and solar-powered lanterns.

It is worth taking a few moments to examine this table carefully, as it illustrates several interesting points. Overall the systems with the least capital costs are the ones giving the most expensive light. Note that the cost per hour for candle lighting is about the same as that for a kerosene hurricane lamp: however, it must be remembered that this analysis is for just *one* candle, giving only 16 lumen, compared to 50 for a kerosene lamp. Hence the cost per klm-hour for candles is the highest of all the methods being compared.

Kerosene pressure lamps have the highest cost per hour (double that of the hurricane

Table 10.5 Economic comparisons of portable lanterns

Type of lantern	Candle	Hurricane lamp	Pressure lamp	Gas lamp	PV lantern
Energy source	Paraffin wax	Kerosene	Kerosene	Butane	Solar
Fuel consumption	1 per 4 hours	0.03 l / h	0.07 l / h	30 g /h	none
Fuel price (US$)	0	0.8 $ / l	0.8 $/l	1 $/kg	0
Light output (lm)	16	50	700	500	320
Capital cost (US$)	0.12	8	55	25	150
Replacement costs	-	lamp	lamp	lamp	*
Lamp replacements (nb of time)	-	3	2	3	-
Non-fuel running costs (US$ / year)	44	10	10	10	0
Annualized LCC (US$)	44	48	109	63	37
Cost US$ per hour	0.03	0.032	0.075	0.043	0.025
Cost US$ per klm-hour	1.88	0.66	0.11	0.09	0.08

General hypothesis: 4 hours of daily use, period of analysis: 12 years

* Similar to Table 10.4 (i.e. battery, US$40 every 3 years; fluorescent tube, US$8 every 3 years; control card US$40 every 6 years)

lamp) due to the large amount of fuel that they use. This will be beyond the financial resources of most rural householders, even though the high light output makes it cheaper in terms of cost per klm-hour. Thus pressure lamps are more often used in bars and businesses than in private homes.

In this analysis the gas lamp has a roughly comparable cost per hour to a hurricane lamp, although like the pressure lamp, its high lumen output makes the cost per klm-hour fairly low. The competitiveness of butane lamps will depend on the price and availability of butane which can vary enormously (US$0.3 to $1.5/kg).

The most interesting case in the analysis is the PV lantern. This has the lowest cost per hour, and the lowest cost per klm-hour, despite having the highest capital cost of all the options being compared. This is due to its exceptionally low running costs and its high light output, and means that it can give great improvements in lighting while still being cheaper than (or comparable to) a standard hurricane lamp, when its costs are annualized.

Figure 10.3 shows the annualized cost of each of the above lighting methods, and breaks down each one to show the relative contributions from capital, replacement and annual costs. It should of course be realized that the results are dependent on the local costs that have been used, and that in another location the balance will be different. This is likely to depend largely on the local costs of kerosene and gas as well as their rate of consumption. In this comparison, figures have been used that are consistent with those elsewhere in the book.

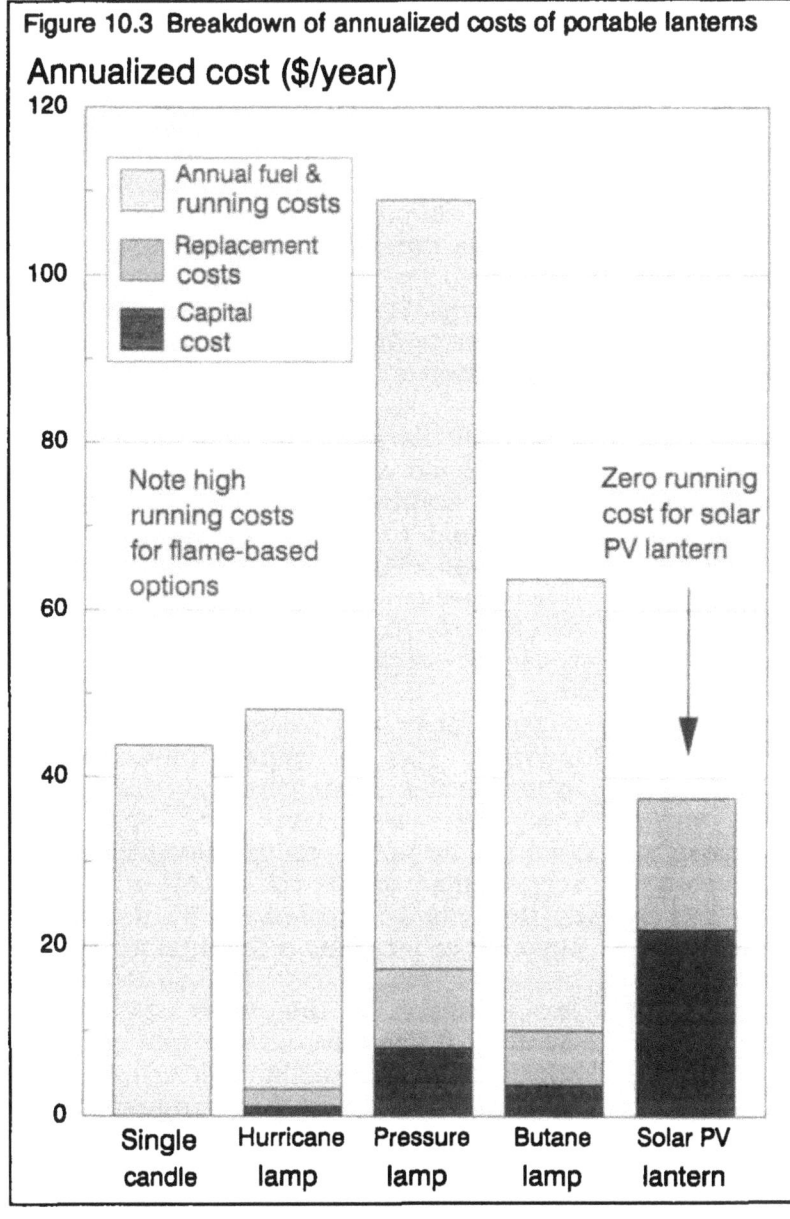

Figure 10.3 Breakdown of annualized costs of portable lanterns

Annualized cost ($/year)

- Annual fuel & running costs
- Replacement costs
- Capital cost

Note high running costs for flame-based options

Zero running cost for solar PV lantern

Single candle | Hurricane lamp | Pressure lamp | Butane lamp | Solar PV lantern

Sensitivity of results to changes in parameters

Throughout the examples above, it can be seen that economics of different types of lighting or power supply systems are dependent on various parameters, and are more sensitive to variations in some quantities than in others. In general, the annualized cost of any system consuming fuel will have greatest sensitivity to fuel price and hours of use, and will usually be less sensitive to the capital cost of the equipment. This is true for kerosene lamps, gas lamps, etc. and also for electrical systems powered by diesel or gasoline (petrol). It also relates to grid-connected systems, where the cost of the electricity per kWh is effectively a fuel cost.

On the other hand, annualized costs of the renewable power sources are more sensitive to the capital cost of the equipment, as this makes up the bulk of the life-cycle cost. For the renewables the capital cost is largely dependent on the sizing of the system (i.e. size of PV array or wind generator). As well as being dependent on the energy demand, the size also depends on the level of the resource available. In the case of PV this is the amount of sunshine, measured in $kWh/m^2/day$ (see Chapter 8). In effect this means that PV is often only economic in tropical or subtropical areas (e.g. with year-round insolations greater than about 4 or 5 $kWh/m^2/day$) as in higher latitudes the lower sunshine levels mean that the system size, and hence the capital cost, is too great.

Similarly for wind-powered systems, the size of turbine necessary depends on the strength of wind available. It is not possible to generalize, because the resource can vary so much from site to site, but the end result is that only in areas with a year-round mean windspeed above a certain value will it be economic to use wind power. Because the power in the wind is proportional to the cube of the windspeed (see Chapter 8) the size, and hence the cost of a wind generator is extremely sensitive to windspeed.

10.5 Financing Lighting Systems

Despite the fact that modern lighting systems (i.e. electrical) can be more economical on a life-cycle cost basis than traditional systems, their capital costs remain very high. This will hinder their diffusion among the majority of rural people and in rural buildings unless some financial assistance is provided.

Ideally, financial assistance for a modern lighting system should allow people to pay for the system without exceeding their present weekly or monthly expenditures on lighting (i.e. fuel). The rationale is that people already spend money on a weekly or monthly basis to maintain a lighting system (from only US$15 up to US$200 per year per household, depending on income level). Most traditional systems, however, are comparatively expensive to run as has been demonstrated in the above sections. Hence, financing mechanisms which can offset the capital costs of more efficient systems and rationalize weekly or monthly household expenditures for lighting, can effectively lower the costs of lighting for a household or business.

For several years now and in more than 20 developing countries, various public and private development projects introducing modern lighting systems have been implemented. These include the introduction of compact fluorescent lamps where grid electricity is available and PV, wind-powered or biogas lighting systems in remote rural areas. Apart from minor technical problems, the major problem is often the same: offsetting the high capital costs, as most people for whom these schemes are designed have limited or no capital resources.

Rural people commonly have limited access to formal credit for various reasons: lack of collateral, high borrower transaction costs, limited formal education (low level of literacy and numeracy) and social and cultural barriers (women may be excluded from local organizations that could provide credit). However, well-designed and implemented savings and credit schemes can be effective to allow rural people to have access to technologies such as modern lighting systems, which in turn can help to alleviate some poverty.

The following sections describe various financing schemes. These are not blueprints but can be adapted according to a country's specific technical, economic and social context. Examples are also provided to give ideas for achieving the introduction of modern lighting systems.

Co-operative or commercial bank loan schemes

Co-operatives and/or commercial banks may provide individual loans on a lighting system for a period ranging from two to five years, with or without an advance payment. The maximum duration of the loan usually does not exceed the lifetime of the major system components likely to fail within this period (e.g. three to five years for batteries in a PV lighting system).

Ideally for the client, the monthly loan payments should be in the range of the previous monthly lighting expenditure. The interest rate should be as low as possible, and if possible subsidized by a development bank. Such schemes have been used in Sri Lanka and Rwanda. The Rwandese project is described in the box that follows.

A variation of this financing scheme is an individual loan for certain components of lighting systems that maintain a high resale value over several years (e.g. photovoltaic modules for a PV lighting system). These components can serve as a form of collateral as they can be repossessed by the lending organization if payments are defaulted.

One advantage of this system is to avoid financing a loan on components having a shorter lifetime than the loan (e.g. batteries for PV or wind-powered lighting systems). As the total loan is smaller, the duration of the loan can be extended (e.g. up to 10 years for PV modules) and therefore the monthly payments will be lower and become eventually more affordable and attractive for more people. This system assumes, however, that the people can afford to pay cash for a small battery, cables and at least one lamp.

Renting schemes

Renting schemes have the advantage of offsetting all capital costs and avoid putting financial contractual burdens (i.e. repayment of loans) onto the users. To implement such a scheme, some initial capital must be made available, for example, by an aid organization, to purchase all the systems. The main drawback of this scheme is the lack of ownership of the systems by the users which may

Commercial banks involvement in financing energy efficient lighting systems in Rwanda

In Rwanda, as in many developing countries, 95% of the population lives in rural, non-electrified areas. The government has launched a costly rural electrification programme The cost of grid extension however is very high: the equivalent of US$3350 for a transformer, US$5600 per kilometre of line, etc. Each individual connection to the grid costs US$400 to the user plus US$80 for each lamp installed.

In comparison, PV systems offer a lower cost option. The initial capital and installation costs of a solar lighting kit comprising four 8W lamps is US$560 and there is no bill at the end of each month. Although this may seem promising, it is of course quite impossible for a Rwandese peasant to find a single payment as large as US$560. Thus the 'Union of Popular Banks' has agreed to provide interested people with appropriate loans in order to cover the expense, provided that:

+ there are at least five interested people in the village;

+ each person puts up 20% of the cost, i.e. US$110.

When these conditions are met, the bank pays the total cost of the system to the supplying company, who then proceed with the installation. The reimbursement of the loans is spread over 12 to 37 months with installments ranging from US$14 to $40 per month. As a result, 1450 lighting kits have been sold and installed in 1991. (See Bibliography, Ngabonziza)

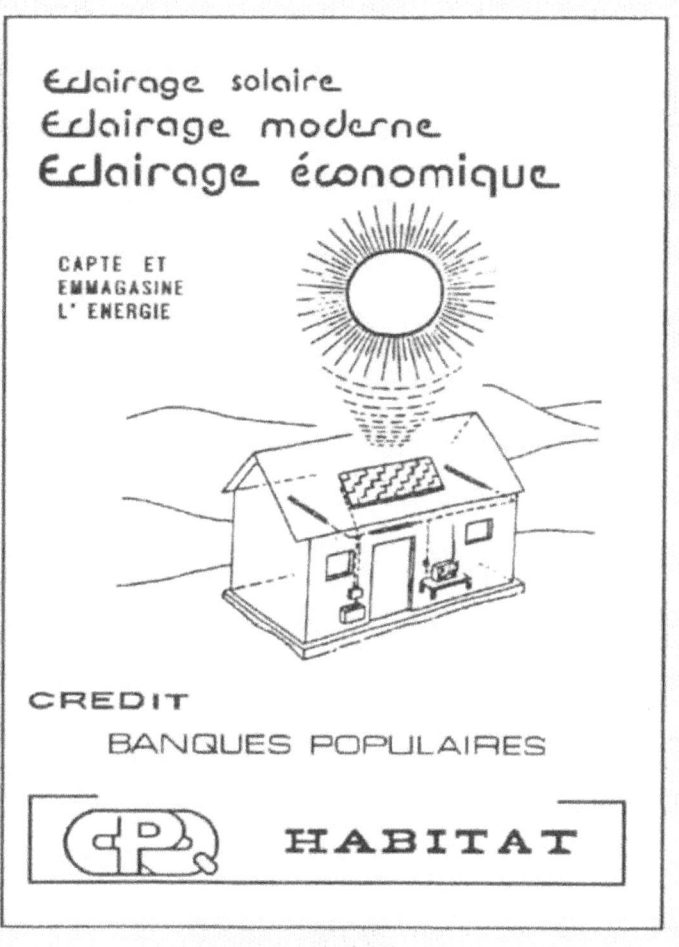

eventually lead to premature failures due to misuse or even theft.

Such schemes have been implemented in various west African countries. In Senegal (since 1991) 36 systems, comprising a PV recharging station with five PV portable lamps each, are being rented. The project was managed at the central level by the Centre d'Etudes et de Recherche sur les Energies Renouvelables in Dakar, and at the local level by existing local organizations. Between 10 and 15 lamps were rented by villagers at a cost of US$0.38 for each recharge. Despite some technical problems due to the poor design of the lamps, the lamps were continually rented for 80% of the time (i.e. 24 days/month), (See Bibliography, Farcot, A.).

Leasing schemes

Leasing schemes can allow the user to rent a lighting system, with an option to buy. In this case, the sense of ownership is often much greater, leading to extended lifetime of the system and making it more economical. Such schemes have been implemented in the Philippines and in Indonesia for the introduction of modern lighting systems in remote areas.

In areas where the grid is overloaded or if the generation capacity is insufficient, electrical utilities will lease CFLs to customers for a small monthly fee (e.g. US$0.25). The lease payments, spread over a few years, will repay the interest-free costs of the CFLs. Each month the CFLs save consumers more in energy bills than they cost in lease payments. Such schemes are being developed in Brazil, Mexico and India.

Partial or total subsidies

Partial or total subsidy schemes can be applied to reduce or fully offset the capital costs of any energy-efficient lighting system. Where the grid is available, utilities may provide compact fluorescent lamps as a replacement for incandescent lamps at a reduced price. For utilities which lack power production capacity, as is the case in many developing countries, it is more profitable to subsidize energy-efficient appliances than to

build new power stations. In some cases, the compact fluorescent lamps are given free of charge, bearing in mind that it is still cheaper from the utilities' point of view and of course cheaper from the user's point of view. Such schemes have been implemented in some parts of UK and in the United States.

A total subsidy can also be provided for the electricity generator (e.g. PV module). The user pays only for the luminaires, batteries, regulator and cables. Such schemes have been successfully implemented in Martinique, Guadeloupe and French Guyana.

Revolving funds

With this scheme, some pioneering users are paying a monthly sum that is free (or almost free) of interest over a long period of time until he or she owns the system. The regular payments are deposited into a fund used to purchase other solar lighting systems which

---

Revolving funds for solar lighting systems in Lebak, Indonesia

Before the project, the inhabitants of Lebak in Indonesia were mainly using kerosene lamps for lighting and batteries for the use of radios and TVs. Villagers were paying between US$0.20 to 0.40 for one litre of kerosene. The total amount of money spent each month was approximately US$5.40 to US$12.00. This amount is now redirected to the loan payments of simple solar lighting systems. These systems were sold to the inhabitants through the local co-operative by means of loans without interest. During 10 years, a villager will pay approximately US$4.00 per month: after which the system belongs to him/her. A small amount is reserved for maintenance and repair costs. Through this system, a financial fund is developed and at the same time it is used to facilitate other systems in the village (or elsewhere). Since the project began in 1991, 500 houses have been equipped with PV lighting systems.

Solar lighting systems in Sukatani, Indonesia installed in 1989. The Sukatani project was the precursor of the Lebak project

will be installed in the community (or elsewhere) as the project continues. In order to begin such a scheme, a donation (seed money) is needed by an aid agency or private company, of money and/or equipment, to install the first systems. Again, the monthly payments should reflect previous monthly expenditures for lighting when conventional systems were used (e.g. kerosene, candles, battery recharging). Thousands of solar lighting systems have been installed using revolving funds in the Dominican Republic (1500 systems), in Morocco (several hundreds) and in Indonesia.

Ideal conditions

Financing opportunities have resulted in thousands of solar lighting systems being installed in developing countries. These schemes have a better chance of success if the following conditions are satisfied:

- National or regional policies towards rural electrification open to cost-effective solutions even if the capital cost is high. (This may have the effect of reducing taxes and/or creating subsidies);

- Local/community institutions with a strong commitment, e.g. development committees, cooperatives or rotating savings and credit associations. Such organisations help to bridge the gap between rural people and the banking system by forging links between them. (see box 'money granaries');

- Availability, on a commercial basis, of a range of reliable, highly efficient and appropriate lighting systems, in a wide price range (e.g. PV portable lamps for people with lower-income circumstances, fixed PV lighting systems for those with higher incomes);

- Long-term technical and financial infrastructure for the maintenance of the systems, (see box 'Income generation to finance services');

- Involvement of the private sector, i.e. local companies, in the installation, training of users and maintenance of the systems, and eventually in the manufacture of various components.

Innovative financing mechanisms

Where several modern lighting systems have been installed, it is necessary and beneficial to have regular maintenance of the systems in order to ensure their longevity and economic benefits. However, financing of a maintenance infrastructure or service presents added costs to the user or the project. One way to defray these costs or to generate additional resources from the systems, is to install additional power generators. Representing a marginal portion of the entire project costs, these generators can sell energy which provides additional resources for the project. The additional power can be sold to the community in the form of a battery recharging service or a secondary cell rental scheme.

The 'money granaries'

'Money granary' is the name given to a novel village banking institution which has been developed in Mali by rural extension agencies. The money granary has been created and managed by the villagers themselves. A literate villager is in charge of the day to day function of the bank. He takes in villagers' deposits and keeps a small amount of money (i.e. just enough to pay the usual withdrawals). The bulk of the savings is deposited in a savings account at the nearest branch of the 'National Agriculture Development Bank' (BNDA). The granary can make loans to any of its members who need them for an important expense, for instance the purchase of agricultural equipment. If the resources of the money granary are insufficient to cover all the demands, then the BNDA will accord it a loan. The experience with all the money granaries created so far has been very successful. Such a system could be adapted to help villagers to afford energy efficient lighting systems.

Income generation to finance services

Since July 1990, several pilot systems for the commercialization of solar energy have operated in Zaire in the Rural Health Zone of Nselo.

One method of redistributing the initial cost of solar energy and recuperating the ongoing costs of the Expanded Programme of Immunizatiom (EPI) is to generate surplus solar energy in a health centre by installing a more powerful generator than normally required to operate the refrigerator and/or lighting systems. This surplus of energy can then be 'sold' to the community. Such was the theory behind the EPI project so that the revenue generated from surplus energy would contribute to a hospital or health centre's ability to autofinance itself.

After a study examining the local market and evaluating the potential clientele, a PV battery charger capable of charging one 12V battery per day; a charger of nickel cadmium batteries (200 nickel cadmium batteries size D) and a TV with a video player were installed at the Nselo Hospital. Similarly, at a health centre 25km from Nselo, a 12V battery charger was installed.

Since then, these systems have functioned without major technical problems, and the demand has increased to be greater than the capability of the system to recharge batteries. Despite a slow start on the rental of nickel cadmium batteries, there is now a faithful clientele who consider the batteries more durable and less expensive than the dry cells they can buy locally.

Advertisement for the recharging station of the hospital

Such a scheme was piloted in a rural health district in Zaire where there was an existing solar powered system and maintenance infrastructure. Added to this were additional generators which sold energy to the local community. The revenues, in this project, contributed to the auto-financing of the district hospital and health zone.

10.6 Quick-reference Lamp Costs

The prices given in Table 10.6 are only indicative. However, they give an idea of the price range between different sources of lighting.

For various reasons, the prices may vary greatly, depending for example on whether the lamps are manufactured in the country of purchase or not. In many countries, including developing countries, there are factories which manufacture standard incandescent bulbs and, to a lesser extent, standard (i.e. 36mm diameter) fluorescent tubes. Where this is the case, the prices are usually fairly low, although unfortunately quality can be low as well.

High-quality halogen incandescent lamps or 26mm high-quality fluorescent tubes are rarely locally manufactured. As a consequence, their prices may be quite high. Discharge lamps such as low-pressure sodium, metal halide, etc., are most often imported from industrialized countries.

Lamp prices also depend on whether the lamps are bought in bulk or in a retail shop.

The quality of the lamp (e.g. lifetime, colour rendering) also has a significant influence on price. For example, the price of fluorescent tubes varies by a factor of three from the worst colour rendering to the best.

The purchase cost is, of course, not the only consideration. Standard incandescent bulbs are very cheap to buy, but very expensive to run. A fluorescent tube or a CFL is several times more expensive to buy while requiring only a fraction of the power needed to run them. This is a tremendous advantage over an incandescent bulb because energy is scarce and expensive, particularly in rural areas.

Table 10.6 Indicative prices for lamps

FLAME-BASED LIGHTING SOURCES

Type of lamp	Fuel	Total unit price (US$)
Candle	Wax	0.15
Hurricane lantern	Kerosene	5.00
Kerosene pressure lamp	Kerosene	50.00
Noorie (kerosene)	Kerosene	50.00
Carbide lamp	Carbide	45.00
Gas lamp	LPG	20.00
Biogas lamp	Biogas (gas)	10.00

ELECTRICAL LIGHTING SOURCES

Type of lamp	Lamp power rating (W)	Lamp price (US$)	Fitting, control[1] gear price (US$)	Total unit price (US$)
Incandescent lamps				
standard	60W / 240V	0.50	1.00	1.50
	100W / 240V	0.50	1.00	1.50
halogen standard	75W / 240V	12.00	3.00	15.00
halogen spotlight	20W / 12V	11.00	2.00	13.00
dichroic halogen (12 / 24V)	50W / 12V	11.00	2.00	13.00
halogen car headlight	55W / 12V	3.00	2.00	5.00
Low-pressure discharge lamps				
fluorescent lamps (by diameter)				
16mm tube	4W	3.00	8.00	11.00
	8W	3.00	8.00	11.00
26mm tube	18W	5.00	12.00	17.00
	36W	6.00	15.00	21.00
	58W	8.00	15.00	23.00
36mm tube	75W	12.00	15.00	27.00

Table10.6 Indicative prices for lamps (continued)

ELECTRICAL LIGHTING SOURCES

Type of lamp	Lamp power rating (W)	Lamp price (US$)	Fitting, control[1] gear price (US$)	Total unit price (US$)
Low-pressure discharge lamps (continued)				
CFLs				
integral, inductive control gear	18W / 240V	16.00	1.00	17.00
integral electronic control gear	9W / 240V	18.00	1.00	19.00
	23W / 240V	22.00	1.00	23.00
modular (no control gear)	7W / 240V	5.00	13.00	18.00
	18W / 240V	11.00	11.00	22.00
sodium low pressure lamps	18W / 240V	30.00	50.00	80.00
	55W /240V	45.00	70.00	115.00
High pressure discharge lamps				
mercury vapour lamps	50W / 240V	20.00	30.00	50.00
	250W / 240V	40.00	60.00	100.00
blended lamps	160W / 240V	50.00	3.00	53.00
metal halide lamps	70W / 240V	60.00	100.00	160.00
	150W / 240V	65.00	105.00	170.00
high-pressure sodium lamps	50W / 240V	50.00	90.00	140.00
	250W / 240V	80.00	140.00	220.00

1 This is considering the simplest luminaire, i.e. including fittings to hold the lamp, and a control gear (ballast, ignitor, etc.) if necessary to start and run the lamp. This does not include a reflector or weather protection, etc.

Abbreviations

A	Ampere		**Li**	Luminous intensity (cd)
AC	Alternative current		**LLC**	Life-cycle cost
ADF	Average daylight factor (%)		**LLF**	Loss light factor (%)
Ah	Amperehour		**Lm**	Lumen
ALCC	Annualized life-cycle cost		**LPG**	Liquefied petroleum gases
BC	Bayonet (fitting for lamp)		**Lx**	Lux
BOS	Balance of system component		**N**	Number
C	Capacity (of a battery) (Ah)		**Nicad**	Nickel Cadmium (batteries or cells)
Ca	Payment occurring annually		**M**	Mega
CCT	Correlated colour temperature		**MKS**	Metre/Kilogram/second (system)
cd	Candela		**P**	Power (W)
CFL	Compact fluorescent lamp		**Pa**	Discount factor for annually recurring costs
Cg	Coefficient for control gear		**Pmin**	Power minimum (W)
CGS	Centimetre Gram Second (system)		**Pr**	Discount factor for a single future payment
CIE	Commission Internationale de l'Eclairage		**Pu**	Unit power (W)
Cr	Single future cost		**PV**	Photovoltaic
CRI	Colour rendering index		**PW**	Present worth (net present worth)
Cu	Usable capacity (Ah)		**s**	Second
C/n	Discharge or charge rate (battery)		**S**	Surface to be lit (m²)
DC	Direct current		**SI**	Système International
DOD	Depth of discharge (%)		**SIL**	Standard incandescent lamp
E	Illuminance (lx)		**SOC**	State of charge (%)
Ep	Practical illuminance (lx)		**UF**	Utilization factor (%)
ES	Edison screw (for lamp fitting)		**UPS**	Uninterruptible power supply
h	Hour		**US$**	United States dollars
HID	High-intensity discharge		**V**	Volt
I	Luminous efficacy (lm/W)		**W**	Watt
Iu	Luminous flux (lm)		**Wh**	Watt-Hour
J	Joule		**Wp**	Watt peak
klm	Kilolumen		**μm**	micrometer (10^{-6}m)
kW	Kilowatt			
kW	Kilowatt-Hour			

Glossary

Cross-referenced terms are shown in italics

Alternating current (AC) - The electric *current* in which the direction of flow is reversed at frequent intervals (50 or 60 times per second). Opposite of *direct current* (DC).

Amorphous - The condition of a solid material in which the atoms are not arranged in an orderly pattern.

Array - Several *photovoltaic modules* connected together.

Average daylight factor (ADF) [unit %] - The extent to which the *illuminance* of an interior is due to *daylight*.

Balance of components (BOS) - Components of a system other than the main ones (e.g. cable fixings of an electrical lighting installation).

Battery - A group of *cells* used to store electrical *energy* (e.g. a car battery).

Beam angle - The angle through the points where the *luminous intensity* is half of its maximum value in an axially symmetrical *light* distribution (e.g. spot light).

Candela - The SI ('Système International') unit of *luminous intensity*, defined as the luminous intensity, in the perpendicular direction, of a surface of 1/600,000 square metre of a black body at the temperature of freezing platinum under a pressure of 101,325 newtons per square metre (i.e. one atmospheric pressure). One candela is equal to one lumen per *steradian*.

Cell - In a *battery*, it is the basic unit that stores electricity. On a *photovoltaic module*, it is the basic unit that produces *direct current* electricity.

Charge rate (C/n) [unit A] - The current at which a *battery* should be charged according to manufacturers' specifications.

Charge/discharge efficiency [unit %] - The ratio (expressed as a percentage) of the amount of *energy* used by the *loads* (e.g. lights) to the amount of energy needed to fully recharge the *battery* over a complete charge/discharge cycle.

Colour rendering - A general expression for the appearance of surface colours when illuminated by *light* from a given source compared with their appearance under light from some reference source, such as *daylight*.

Colour rendering index (CRI) - A measure of the *colour rendering* that is the degree to which the colours of surfaces illuminated by a given *light* source conform to those of the same surfaces under a reference light, such as *daylight*.

Colour temperature [unit K] - The temperature of a full radiator which emits radiation of the same chromaticity as the radiator being considered. In practice, it describes how a *light* source appears (e.g. warm or cold).

Contrast - Differences of appearance between two parts of a visual area (e.g. difference of *luminances*).

Control gear - Electrical devices (e.g. ballast) necessary to start and maintain adequate running of *discharge lamps*, and increase the *power factor* (e.g. ballast).

Correlated colour temperature [unit °K] - The temperature of a full radiator which emits radiation having a chromaticity nearest to that of the *light* source being considered, e.g. the colour of a full radiator at 3500°K is the nearest match to that of a 'white' tubular fluorescent lamp.

Crystalline silicon - Silicon that has solidified at such rate that small crystals have formed. Mono-crystalline silicon have crystals symmetrically ordered while in poly-crystalline, crystals are jumbled together.

Current (electrical) [unit A] - The current, measured in amperes, is the flow of electricity occurring through a conductor due to a *voltage* difference at its ends (an electrical current is analogous to water flow rate created by a difference of height or pressure).

Cut-off voltage [unit V] - Threshold *voltage* after which a *battery* gives no more power without risking a serious reduction of its life.

Cycle life - The number of times (i.e. cycles) a *battery* can be charged and discharged before it permanently loses more than 20% of its *rated capacity*.

Daylight - Daylight is a combination of *skylight* and any available *sunlight*.

Depth of discharge (DOD) [unit %] - The ratio of the *usable capacity* to the *rated capacity* of the *battery*. So a low DOD implies that less charge is taken from the battery at each cycle.

Design service illuminance [unit lx] - Value of *illuminance* suitable for a given task and chosen in the design process of a lighting system.

Diffuse reflection - Reflection in which the reflected *light* is diffused and there is no significant *specular* reflection, as from a matt paint.

Direct current (DC) - Electric *current* in which electrons are flowing in one direction.

Direct lighting - Lighting design where *luminaires* direct 90 to 100% of their light towards the *working plane*.

Discharge lamp - A *lamp* in which *light* is produced by an electric discharge through a gas, a metal vapour or through a mixture of both.

Discharge rate (C/n) [unit A] - The discharge *current* of a *battery* which is expressed as a fraction of the *rated capacity*.

Discount factor - A factor that when multiplied by future costs or benefits, gives their *present worth*. Calculated from the *discount rate*, inflation relative to the general rate, and the number of years in the future that the payment will be made.

Discount rate - The annual rate at which the *present worth* of future costs or benefits decreases. Also known as the opportunity cost of capital. Around 10 to 12% for most economies.

Effective reflectance [unit %] - Estimated reflectance of a surface, based on the relative areas and the reflectances of the materials forming the surfaces.

Electrical capacity [unit Ah] - Quantity of electrical *energy* contained in a *battery* expressed in ampere-hour (Ah).

Electrical light source - Any light source that needs electricity to produce light. They can be divided into two broad classes on the basis of their operating principles: *incandescent lamps* and *discharge lamps*.

Energy [unit J] - Energy is the product of the mean *power* (W) multiplied by the time (seconds). It is expressed in Joule (1 Joule = 1 Watt x 1 second).

Flame-based light source - Any *light* source that produces light directly by a flame.

Flicker - Fluctuation in the *light* output.

Fluorescent lamp - A *discharge lamp* in which most of the light is produced by a layer of phosphorescent material, coated inside the glass, excited by the ultraviolet radiation from the discharge.

Flux - The rate of transfer of a fluid, particules or energy across a given surface.

General lighting - Lighting designed to illuminate an entire area uniformly, without provision for special requirements or activities.

Glare - The discomfort or impairment of vision experienced when parts of the visual field are excessively bright in relation to the general surroundings.

Heating value [unit J/l or J/kg] - The *energy* content of a fuel is expressed by its heating value, either its 'net' or 'gross' heating value. The 'gross' equals the total heat released by combustion while the 'net' excludes the latent heat from combustion gas condensates.

Hydrometer - Apparatus used for the measurement of specific gravity (of the electrolyte in a lead-acid *battery*).

Illuminance (E) [unit lx] - The illuminance is the *luminous flux* falling on a unit area of a surface, and is expressed in *lux* (*lumen* per square metre). In practice, it indicates how well an illuminated object is lit.

Incandescent lamp - A lamp (e.g. standard bulb) in which *light* is produced by a filament heated to a very high temperature (i.e. to incandescence).

Indirect lighting - Lighting in which the greater part of the *luminous flux* reaches a surface through reflection from other surfaces (e.g. when *luminaires* direct 90 to 100% of their output upward to the ceiling).

Initial efficacy [unit lm/watt] - The *luminous efficacy* when a *lamp* is new (e.g. after 100 hours of first operation).

Initial light output [unit lm] - The *luminous flux* from a *lamp* after approximately 100 hours of operation.

Insolation or solar irradiation [unit kWh/ m² / day] - The *energy* received from the sun per unit area on a daily basis.

Inverter - Device that converts *direct current* to *alternating current*.

Irradiance [unit W/m²] - The flux density on a surface, i.e. the radiant flux incident per unit area of surface (e.g. solar irradiance is 1000W/m² at full sunshine).

Lamp - For electrical light sources, a lamp is the component in a *luminaire* where the light is effectively produced (e.g. an incandescent bulb). For flame-based light sources, a 'lamp' is the complete device that produces light (e.g. kerosene lamp, gas lamp).

Lamp circuit luminous efficacy (unit lm/ circuit W) - The ratio of the *luminous flux* emitted by an electrical *lamp* to the power of the lamp including the power consumption of the *control gear* (if any).

Level of illuminance [unit lx] - similar to *illuminance*. In practice, it indicates how well an illuminated object is lit.

Light - Light is the visible radiation, evaluated according to the sensitivity of the human eye.

Life-cycle costs (LCC) - The lifetime costs associated with a system expressed in terms of present monetary sums (i.e. today's money).

Life-cycle cost analysis - An analysis of a project in monetary terms that takes into account all the future costs and benefits over the life-cycle of the project.

Light loss factor (LLF) - The ratio of the *illuminance* provided by a lighting installation at a stated time, with respect to the initial illuminance (i.e. after approximately 100 hours of operation).

Load - Any device or appliance that is using *power* (or *energy*) for example an incandescent lamp.

Local lighting - Lighting designed to provide the required *illuminance* only over the small area occupied by the task and its immediate surroundings, for example a lamp placed on a desk for reading.

Localized lighting - Lighting designed to provide the required *illuminance* on work areas, together with a lower illuminance for other areas.

Lumen (lm) - The SI unit of *luminous flux* which describes a quantity of *light* emitted by a source. The lumen is the luminous flux emitted within unit solid angle of one *steradian* by a point source having a *luminous intensity* of one *candela*. (1 lm = 1 cd st).

Luminaire - A complete lighting unit comprising one or several *lamps* (i.e. bulb), reflector, *control gear* and fittings.

Luminance [unit cd/m^2] - The *luminous intensity* per unit of area of a *light* source or of an illuminated area. In practice, it indicates how bright is a source.

Luminous efficacy (I) [unit lm/W] - The ratio of the *luminous flux* emitted by a *lamp* to the *power* consumed by the lamp. It is the *energy* to *light* conversion.

Luminous flux (Iu) [unit lm] - The *light* emitted by a source, or received by a surface. The quantity is derived from the *radiation spectrum* of a source by evaluating the radiation in accordance with the *spectral sensitivity* of the standard human eye.

Luminous intensity (Li) [unit cd] - The *luminous flux* for a *light* source per unit of solid angle (or steradian) in a given direction. In practice, it indicates how much light is emitted in one direction. It is expressed in the SI *candela* (cd).

Lux (lx) - The SI unit of *illuminance*. One lux is equal to an illuminance of one *lumen* per square metre (1 lx = 1 lm/m^2).

Luxmeter - An instrument to measure *illuminance* on areas.

Monochromatic - A qualification of *light* with a single pure colour (i.e. only one wavelength).

Multimeter - An apparatus which measures *current*, *voltage*, electrical *resistance* in both *direct current* and *alternating current*.

Open voltage [unit V] - *Voltage* measure at a generating system when no *current* is delivered.

Parallel connection - A method of interconnecting two or more electrical-producing or power-using devices, where the *voltage* produced or required, is not increased, but the current is additive. Opposite to *series connection*.

Payback period - The number of years necessary for the income (or value of benefits) from a project to equal the initial capital cost.

Photovoltaic - Relating to the direct conversion of *daylight* into electricity.

Photovoltaic module - A number of *photovoltaic* cells electrically interconnected and mounted together, usually in a common sealed unit of convenient size for shipping, handling and assembling into *arrays*.

Power [unit W] - The rate of the use of *energy* (over time); (1 Watt = 1 Joule/second).

Power (electrical) [unit W] - The power is equal to the *current* in ampere (A) by the voltage in volt (V); (1 Watt = 1A x 1V).

Power factor - The power factor characterises the difference of phase between *current* and *voltage* in *alternating current* supply. If the *load* includes components that are capacitive or inductive, the peak current does not occur simultaneously with the peak *voltage*. The more out of phase the current and voltage, the smaller the power factor and the greater the power loss in the transmission/distribution lines concerned.

Pre-electrification - Where full scale electrification is impractical for technical or economical reasons, pre-electrification consists of providing a community with a minimal quantity of electricity sufficient for a few low power and efficient *lamps*, radios and other few low power uses. For example, this small quantity of electricity can be provided by a small generator set, a *photovoltaic* installation or a small wind turbine generator.

Present worth - The value of a future cost or benefit in current terms, after being adjusted for future changes in the value of money.

Primary cell - Dry *batteries* (e.g. disposable torch batteries).

Radiation spectrum - A graph that gives the wavelength distribution of the emitted *light* of a light source.

Rare-earth incandescent lamps - *Flame-based* lamps that use rare-earth compounds (e.g. thorium oxide). These compounds fixed on a mantle emit visible radiation when heated to high temperature by an invisible or bluish flame (e.g. in a gas lamp).

Rated capacity [unit Ah] - A *battery's* rated capacity is the total quantity of electrical charge (i.e. *current* x time) in Ah (ampere-hours) that can be drawn from a fully charged state at a specified *discharge rate* and electrolyte temperature before the *voltage* falls to a specified cut-off *voltage*.

Reflectance (average) - The ratio of the *luminous flux* reflected from a plane to the luminous flux falling on it. The value is always less than one and is expressed as either a decimal or a percentage.

Reflection factor - The ratio of the *luminous flux* reflected from a material to the luminous flux falling on it. The value is always less than one and is expressed as either a decimal or a percentage.

Resistance (electrical) [unit Ohm] - The resistance of a circuit is a measure of how well electricity is being conducted (e.g. a copper cable has a low resistance, while most plastic materials have a very high resistance). Resistance is measured in ohms and is equal to the potential difference (or *voltage* drop) divided by the *current*.

Secondary battery - rechargeable *battery*.

Self-discharge [unit %/month] - The self-discharge of a *battery* is the charge lost expressed as a percentage of the initial *state of charge* when the battery is not used over a period of one month.

Series connection - A method of interconnecting two or more electrical-producing or power-using devices, such that the *current* produced or required, is not increased, but the *voltage* is additive. Opposite to *parallel connection*.

Sine-wave - Shape of the *voltage* and *current* from an *alternating current* power source.

Skylight - The scattered *light* received from the luminous parts of the sky, excluding direct *sunlight*. Skylight contributes to the *daylight*.

Spectral sensitivity (curve) - A graph showing how electromagnetic radiation, for example visible radiation, is perceived by the human eye.

Specular reflection - Reflection without *diffusion* in accordance with the laws of optical reflection as in a mirror.

State of charge (SOC) [unit %] - The amount of charge (i.e. electricity) left in a *battery* expressed as a percentage of the *rated capacity*.

Steradian (st) - Unit of solid angle measurement. A steradian can be thought of as follows: on the area of a sphere with a radius of one metre, imagine an area of one square metre; the shape of this area is immaterial. If a radius of the sphere is allowed to follow the perimeter of this area, then this radius describes a cone which encloses a solid angle unit; this is one steradian.

Stroboscopic effect - An illusion caused by oscillation in *luminous flux*, that makes a moving object appear stationary or moving in a manner different from how it is truly moving.

Sunlight - The *light* received directly from the sun. Sunlight contributes to *daylight* along with *skylight*.

Task lighting - similar to *Local lighting*.

Transmittance - The ratio of the *luminous flux* transmitted by a material to the incident flux.

Usable capacity (Cu) [unit Ah] - This is the fraction of the *rated capacity* of a *battery* that is used for each cycle (to allow a reasonable *cycle life*).

Utilization factor (UF) [unit %] - The proportion of the *luminous flux* emitted by a *lamp* which reaches the *working plane*.

Voltage [unit V] - The measure of the electrical potential (i.e. voltage) difference between two conductors or one conductor and the earth. It is expressed in volts (V).

Watt peak [Unit Wp] - The approximate amount of *power* that a *photovoltaic* module or array produces at noon on a clear day, such that the *irradiance* is equal to 1000W per square metre when the device directly faces the sun.

Working plane - The horizontal, vertical or inclined plane in which the visual task lies. Unless otherwise specified, the working plane may be considered to be horizontal and at 0.7m to 0.85m above the floor (for example, a table).

Bibliography

Code for Interior Lighting 1984, CIBSE, Delta House, 222 Balham High Road, London SW12 9BS, UK.

Derrick, A., Francis, C., and Bokalders, V., *Solar Photovoltaic Products, A Guide for Development Workers*, IT Publications, London, UK, 1992. ISBN 1-85339 091-7

Farcot, A. and Efforsat, J., 'Les lampes portables,' *Solar for Health*, WHO/EPI/SOL/WP.1, World Solar Summit, July 5-9,1993.

Gadgil, A., *Stalled on the Road to the Market: Analysis of Filed Experience with a Project to Promote Lighting Efficiency in India*, Proceedings, Vol. 6, pp. 6.57-6.70, ACEEE 1992 Summer Study on Energy Efficiency in Buildings, Asilomar, CA.

Gadgil, A. and Jannuzzi, G., 'Conservation Potential of Compact Fluorescent Lamps in India and Brazil,' *Energy Policy*, Volume 19, No. 5, June 1991, pp. 449-463.

Gadgil, A. *et al.*, *Advanced lighting and window technologies for reducing electricity consumption and peak demand: overseas manufacturing and marketing opportunities*, LBL-30389, proceedings IEA/ENEL Conference on Advance Technologies for electric Demand-side Management, Italy, April 4-5, 1991.

GTZ, *Biofuels for developing countries: promising strategy or dead end?*, GTZ GmbH, Dag-Hammarskjold 1/2, D 6236 Eschborn 1, Germany, 1985.

Hankins, M., *Small solar electric systems for Africa*, Commonwealth Science Council, London, Motif Creative Arts Ltd, Kenya, May 1991. ISBN 0-85092 374-3

Leach, G. and Gowen, M., *Household Energy Handbook*, World Bank Technical Paper No. 67, Washington D.C., 1987.

Louineau, J-P. 'Le Solaire C'est la Santé,' *Systèmes Solaires*, No. 73, Comité d'Action pour le Solaire, Paris, France, Nov. 1991.

Ngabonziza, J-D-D. 'Leasing solar energy,' *Best of Systèmes Solaires*, No. 1 Comité d'Action pour le Solaire, Paris, France, April, 1993.

Rajvanshi, K. and Kumar, S., *Development of Improved Lanterns for Rural Areas*, Nimbkar Agricultural Research Institute (NARI), Phaltan 415 523 Maharashtra, India, 1989.

Rajvanshi, K., and Kumar, S., *Ethanol from sweet sorghum*, Nimbkar Agricultural Research Institute (NARI), Phaltan 415 523 Maharashtra, India, 1989.

Roberts, S., Solar electricity, A practical Guide to Designing and Installing Small Photovoltaic Systems, Prentice Hall International, Hertfordshire, UK, 1991. ISBN 0-13-826314-0

Van der Paas, R. and de Graaf, A., *A comparison of lamps for domestic lighting in developing countries*, Industry and Energy Department Working Paper, energies series No. 6, World Bank, Washington D.C., USA, June 1988.

WHO, *Logistics for Health Series: Sale of Excess Solar Energy*, WHO/EPI/LHIS/90.1, WHO, Geneva, Switzerland, 1990.

APPENDIX A

Address list of manufacturers and system suppliers

The list of manufacturers of lighting systems is divided into five sections:

- Electrical lighting;
- Solar lighting;
- Flame-based lighting;
- Small wind electricity generators;
- Miniature hydro-electric turbines.

The electrical lighting section consists of international manufacturers of bulbs, lamps and luminaires of all kinds. This section is short because even in developing countries, it is relatively easy to find what electrical lighting systems are available just by going to large local firms in the capital city or large towns which are electrified.

In most cases, these companies import the lighting devices from the companies mentioned in this list. If catalogues are required, the address list gives contact details of the major multinational companies, whose products include most of the world's well-known electrical lighting systems.

The second section concerns solar lighting systems. It is larger because the technology is less well-known but has great potential. The major western companies are listed as well as the growing numbers of local firms in the developing world, although it is impossible to give a comprehensive list of companies in the developing world. It is suggested that this list is used as a starting point and when in a particular country look for other companies involved in the field.

The list for flame-based lighting systems is quite short, as these products include kerosene lamps, candles and even the custom-made 'tomato-tin oil lamp'. They are either widely available in rural markets even at the village level, for example, Chinese-made kerosene wick lamps, or are manufactured locally, for example some types of biogas lamp.

Finally, a list of manufacturers of small wind turbines and miniature hydro-electric turbines is provided as such systems may be suitable to generate small quantities of electricity for lighting in remote areas not connected to the electricity grid.

ELECTRICAL LIGHTING

General Electric
Nela Park
Cleveland
Ohio 44112
USA
tel +1 216 266 2653
fax +1 216 266 2371

GTE Sylvania Ltd
Otley Road
Charleston, West Yorshire
BD17 7SN
United Kingdom
tel +44 274 59 59 21
fax +44 274 59 76 83
tlx 51251

Harvey Hubbel Lighting
Ronald Close
Kempston, Bedford
MK42 7SH
United Kingdom
tel +44 234 85 54 44
fax +44 234 85 40 08
tlx 826065

International Lamps Ltd
South St
Herrford, Herts
SG14 1BL
United Kingdom
tel +44 992 55 44 11
fax +44 992 58 77 66
tlx 817296 LAMPS G

Lab-Craft Ltd
Bilton Rd, Waterhouse Lane
Chelsford, Essex
CM1 2UP
United Kingdom
tel +44 245 35 98 88
fax +44 245 49 07 24
tlx 995291 LABSLS G

Mazda Eclairage
Tour Chenonceaux, Pt de Sèvres
Boulogne Billancourt
F-92516
France
tel +33 1 46 20 79 21
fax +33 1 46 20 76 54
tlx 631 484

Moorlite Electrical Ltd
Burlington Street
Ashton-Under-Lyne, Lancashire
OL7 0AX
United Kingdom
tel +44 61 330 68 11
fax +44 61 330 28 15
tlx 668284

Osram Gmbh
-
81543
Munich
Germany
tel +49 896 21 31
fax +49 896 21 32020

Philips Lighting
BV Building ED1
PO Box 80020
5600 JM, Eindhoven
The Netherlands
tel +31 40 75 70 65
fax +31 40 75 54 53
tlx 35000 PHTC-NL

Progressive Technology in Lighting
581 Ottawa Ave.
Holland, MI
49423
USA
tel +1 616 396 2556
fax +1 616 396 0686

R&S Components
PO Box 99
Corby, Northants
NN17 9RS
United Kingdom
tel +44 536 201 234
fax +44 536 201 501
tlx 42512

Silverlight
Maidstone Rd, Matfield
Tonbridge, Kent
TN12 7JN
United Kingdom
tel +44 892 72 22 02
fax +44 892 72 35 07
tlx 95539 BERK G

ELECTRICAL LIGHTING cont.

Thin-lite Corp.
530 Constitution Avenue
Camarrillo, CA
93012-8595
USA
tel +1 805 987 5021
fax +1 805 388 0921
tlx 662203

Thorn Europhane
156 Bd Hausseman
Paris, cedex
F-75379
France
tel +33 1 49 53 62 62
fax +33 1 49 53 62 10
tlx 651 061

Verre & Quartz
24 rue d'Aulnay
Bondy
F-93147
France
tel +33 48 48 14 22
fax +33 48 48 68 18
tlx 233593F

SOLAR LIGHTING   AFRICA

Agriconsult (T) Ltd
PO Box 74272
Dar Es Salaam
Tanzania
tel +255 51 31 600
fax +255 51 31 136

Casabloc
163 rue Hadj Amar Riffi
Casablanca,1
Morocco
tel +212 31 81 40
fax +212 31 80 41

**Criterion/Gold River
Enterprise**
PO Box 4839, Accra
Ghana
tel +233 21 22 65 44
fax +233 21 22 18 50

Enesol
PO Box 520
Butare
Rwanda
tel +250 30 702

FNMA
14me Rue Limete B.P. 1967
Kinshasa,1
Zaire
tel +243 12 77 264
tlx 20080

Helios Power (Pty) Ltd
Gateway Industrial Park
Pretoria
South Africa

Intertec (Tanzania) Ltd
PO Box 40365 Pugu Road
Dar Es Salaam
Tanzania
tel +255 51 62 361
fax +255 51 62 080

Kenital Solar Electricity
Elgeyo Marakwet Close 381
Nairobi
Kenya
fax +254 2 562 295

Kludjeson International Ltd
PO Box 10011, Airport
Accra
Ghana
tel +233 21 77 21 85
fax +233 21 77 27 37

National Luna Products (Pty)
cnr Harris Ave & Wagenaar Rd
PO Box 8899, Edenglen
ZA-1613, Johannesburg
South Africa
tel +27 011 452 5438
fax +27 011 452 5263
tlx 7-44904

Nord-Industrie
26 Av. kheireddine Pacha
Tunis
Tunisia
tel +216 1 288 746

Plein Soleil
Km 5, Route de Rufisque
Dakar
Senegal
tel +221 32 93 59

Scan African Trading Ltd
PO Box 40490
Gabarone
Botswana
tel +267 313 638
tlx 2638 SCANT BD

Solar Age Namibia
PO Box 9987
Windhoek, N17
Namibia

Solarcomm
12 Lobengula Close
PO Box ST 319, Southerton
Harare
Zimbabwe
tel +263 4 643 41/2
fax +263 4 618 81

Somafrec
Rue Enseigne Froger
BP800
Bamako
Mali
tel +223 225 584

West African Batteries
16 Keffi Street
S/W Ikoyi, PO Box 2341
Lagos
Nigeria
tel +234 1 685 095
fax +234 1 685 182

Bharat Heavy Electricals Ltd
Mysore Road, PO No. 2606
Bangalore
IN-560 026
India
tel +91 812 60 30 24
fax +91 812 62 31 37

Central Electronics Ltd
4 Industrial Area
Sahibad
IN-201 010
India
tel +91 11 8 610 65
fax +91 11 8 73 19 41

Daming Solar Cell Co Ltd
West Bldg, 201,5/F Shang Bu
Ind. District, Shenzhen
518045, Guangdong
P.R. China
tel +86 755 36 51 97
fax +86 755 36 51 97

Hoxan Corporation
13-12 Ginza, 5-Chome Chuo-ku
Tokyo
104
Japan
tel +81 3 3543 2017
fax +81 3 3546 1637

Kyocera Corporation
Karasuma office
680 Karasuma Bukkoji-Sagaru
Shimogyo-ku, Kyoto, 600
Japan
tel +81 75 344 82 41
fax +81 75 344 82 40

Power & Sun Private Ltd
338 T.B.Jayah Mawatha
Colombo
10
Sri Lanka
tel +94 168 63 07
fax +94 1 57 55 99

Qinhuangdao Huamei Photo-Voltaic Electronic Co. Ltd
No. 6 Jianguo Road
Qinhuangdao, Hebei Province
066000
P.R. China
tel +86 532 33 12 34

Rajasthan Electronics Instruments Ltd (REIL)
D-37 Madho Singh Rd, Bani Pk.
Jaipur
302006
India
tel +91 141 62 601

Solar Lab
Trung Tam Vat Ly
01 Mac Dinh Chi
Ho Chi Minh City, Q1. TP
Q1. TP
Vietnam
tel +84 22 028

151

Solar Power Division
PT Kemenangam, JL Gunung
Sahari 75, Jakarta
PO Box 2628
Indonesia
tel +62 21 420 08 23
fax +62 21 420 00 52

Sunergy Pte Ltd
No 12, Lorong Bakar Batu
Kolam Ayer Industrial Estate
1334
Singapore
tel +65 748 81 01
tlx   RS 39 841

Suryovonics Ltd
7-1-21A Begumpet
Hyderbad
IN-500 016
India
tel +91 842 33 827
fax +91 842 41 022

Tata BP Solar India Ltd
A101 Block II, KSSIDC Multi-
storey Blks, Hebbagodi,Hosur
Bangalore, IN-562 158
India
tel +91 812 42 20 83
fax +91 812 42 24 17
tlx   08408 224 TABP IN

Wisdom Light Group (P) Ltd
PO Box 1601
Durbar Marg
Kathmandu
Nepal
tel +977 1 22 86 96
fax +977 1 22 20 26

**Yunnan Semiconductor
Device Factory**
24 Jianshe Road
Kunming City, Yunnam Province
650033
P.R. China
tel +86 871 563 45

Elante Pty Ltd
382 Canterbury Road
Surrey Hills, Vic
3127
Australia
tel +61 3 836 99 66
fax +61 3 836 67 43

Going Solar
320 Victoria Street
North Melbourne
3051
Australia
tel +61 3 328 41 23

Suntron Energy Company Pty
Unit 2, 66-70 Railway Road
Blackburn, Vic
3130
Australia
tel +61 3 894 25 44
fax +61 3 894 33 70

Audiolight A/S
Ensjoveien 20A
Oslo 6
Norway
tel +47 2 19 42 00
fax +47 2 68 00 59

BP Solar Ltd
Solar House, Bridge Street
Leatherhead, Surrey
KT22 8BZ
United Kingdom
tel +44 37 237 78 99
fax +44 372 37 77 50
tlx   263220 BPSIL

Cristina Mejias Romero(CMR)
Lomo del Capon 74 Tafira Alta
Las Palmas (GC)
E-35017
Spain
fax +34 28 35 53 03

Dulas Engineering
The Old School
Eglwysfach, Machynlleth
Powys SY20 8SX
United Kingdom
tel +44 654 78 13 32
fax +44 654 78 13 90

Ecosolaire Serelio
19 rue Pavée
Paris
F-75004
France
tel +33 1 48 87 43 60
fax +33 1 48 87 86 27
tlx   214 235 F

Ecotecnia S.C.C.L.
Demostenes 6
Barcelona
ES-08028
Spain
tel +34 93 330 78 60
fax +34 93 411 23 45

**Energies Nouvelles &
Environnement (ENE)**
Av. Van der Meerschen 188
Brussels
B-1150
Belgium
tel +32 27 71 13 28
fax +32 27 71 13 28

Engcotec GmbH
Konigstras. 72, PO Box 101262
Stuttgart 10
D-7000
Germany
tel +49 711 29 93 61/64
fax +49 711 29 93 66
tlx   721 886

Gallivare Photovoltaic AB
Foretagscentrum, Box 840
Gallivare
S-98228
Sweden
tel +46 970 168 45
fax +46 970 158 98

Helios Technology Spa
Via Postumia 11
Carmignano di Brenta/PD
I-35010
Italy
tel +39 49 943 02 88
fax +39 49 595 82 55

Hiltec Solar Ltd
25 Low Friar Street
Newcastle upon Tyne
NE1 5UE
United Kingdom
tel +44 91 232 88 18
fax +44 91 261 57 46

Inpro-Solar-Systems
Miesberg 2
Odelzhausen
G-8063
Germany
tel +49 081 34 60 24
fax +49 081 34 64 02

**International Battery & Solar
Power Company (IBC)**
Am Hochgericht 15
Staffelstein, Postfach 1107
D-8623
Germany
tel +49 9573 30 66
fax +49 9573 312 64

Intersolar Ltd
Unit 2, Cock Lane
High Wycombe, Bucks
HP13 7DE
United Kingdom
tel +44 494 45 29 45
fax +44 494 43 70 45
tlx   837 383 SOLPAKG

Intertechnology
Bolshoi Volokolamskii
Proezd 16, PO Box 12
Moscow 123436
Russia
tel +7 95 194 82 76
fax +7 95-194 30 60

Isofoton
C/. Miguel Angel 16
Madrid
E-28010
Spain
tel +34 410 23 54
fax +34 410 59 89
tlx   49944 ISOF E

Koncar Solar Cells
Tezacki put bb
Split
CR-58000
Croatia
tel +38 3858 52 54 22
fax +38 3858 51 21 38

Muntwyler Energietechnick
Maeritgasse 1, Ziegelei-Marit
Postfach 512, Zollikofen
CH-3052
Switzerland
tel +41 31 57 50 63
fax +41 31 57 51 27

Neste Advanced Power Systems (NAPS)
Rälssitie 7, Vantaa
SF-01510
Finland
tel +358 0 450 57 53
fax +358 0 826 301
tlx   124641 NESTE SF

Photowatt International S.A.
33 r. St Honoré, Z.I Champfleuri
Bourgoin-Jallieu
F-38300
France
tel +33 74 93 80 20
fax +33 74 93 80 40
tlx   308 270 F

R & S Renewable Energy Energy Systems BV
Lagedijk 26, PO Box 3049
5700 JC Helmond
The Netherlands
tel +31 49 20 23 335
fax +31 49 20 49 665
tlx   59030 RES NL

Siemens Solar GmbH
Frankfurter Ring 152
Munich 40
D-8000
Germany
tel +49 89 3500 24 11
fax +49 89 35 00 25 73

Solar Electric Light Company BV (SELC)
PO Box 45, Eindhoven
NL-5600 AA
The Netherlands
tel +31 40 52 54 65
fax +31 40 55 06 25

Solar Energie-Technik GmbH
Industriestrasse 1 - 3
Postfach 1180, Altlussheim
D-6822
Germany
tel +49 6205 35 25
fax +49 6205 35 28

Solar Products International
PO Box 438
Harrow, Middlesex
HA2 9UT
United Kingdom
tel +44 81 868 83 53
fax +44 81 429 4270

Solart Systems Ltd
Gulyas u. 20
Budapest
H-1112
Hungary
tel +36 1 165 30 37
fax +36 1 165 30 37

Soltech Ltd
Kapeldreef 75
Leuven
B-3030
Belgium
tel +32 16 270 442
fax +32 16 270 443
tlx   26152 (via Imec Leuven)

Steepler Solar
Prechistenka 40
Moscow 119034
Russia
tel +7 95 939 26 82
fax +7 95 939 58 87

Total Energie
7 chemin du Plateau
Lyon-Dardilly
F-69570
France
tel +33 7847 44 55
fax +33 78 64 91 00
tlx   306 115 F

Vergnet S.A.
6 rue Henri Dunant
Ingre
F-45140
France
tel +33 38 43 36 52
fax +33 38 88 30 50
tlx   780 980F

Backwoods Solar Electric Systems
8530 Rapid lightning Creek Rd
Sandpoint, Idaho
83864
USA
tel +1 208 263 42 90

Photocomm, Inc.
7681 East Gray Road
Scottsdale, AZ-85260
USA
tel +1 602 948 80 03
fax +1 602 483 64 31
tlx   823165

Real Goods
966 Mazzoni Street
Ukiah, CA
95482-3471
USA
tel +1 800 762 7325
fax +1 707 468 0301

Remote Power, Inc
1608 Riverside Avenue
Fort Collins, CO
80524
USA
tel +1 303 482 95 07
fax +1 303 482 96 10

Solarex Corporation
630 Solarex Court
Frederick, MD
21701
USA
tel +1 301 998 42 00
fax +1 301 698 42 01

Solec International, Inc.
12533 Chadron Ave
Hawthorne, CA-90250
USA
tel +1 301 970 00 65
fax +1 301 970 10 65
tlx   910 325 6215 SOLEC HWTH

Soltek Solar Energy Ltd
4493 Boundary Road
Vancouver, B.C.
V5N 2N3
Canada
tel +1 604 43 17 66
fax +1 604 43 17 331

Sunnyside Solar, Inc.
Rd 4, PO Box 808
Green river rd., Brattlerbro, VT
05301
USA
tel +1 802 257 14 82
fax +1 802 254 46 70

Empresa de Comunicaciones
Altos de Santa Dom., Las Sie.'s
de la Embajada de Venez
2C al Oeste
Managua, Apartado No. 5547
Nicaragua
tel +502 2 67 57 47
fax +502 2 67 43 87

Heliodinamica S.A.
Rodovia Raposo Tavares
Km 41, Caixa Postal 111
Vargem Grande, Paulista-SP
06730
Brazil
tel +55 11 790 0888
fax +55 11 790 12 80

Heliotecnia S.A.
Apartado de Correos 60699
Chacoa, Caracas
1060
Venezuela

Lacetal
Apartado Postal 6100
La Habana
Cuba
tel +53 7 61 15 31

Mirosolar Chile Etda
Sociedad de Energie Solar
Morambe
855
Chile
tel +56 2 672 21 03
fax +56 2 671 57 23

Representaciones Real S.A.
Apartado 214, Avenida Central
Calles 19-21
San Jose (Y Griega)
1011
Costa Rica
tel +506 22 26 56
fax +506 33 91 61

Soimpro LTDA (Madeco Solar)
Apoquindo No.4823
Santiago
Chile
tel +56 2 228 45 69
fax +56 2 206 13 21

Camping Gaz
173 rue Bercy
Paris
F-75012
France
tel +33 1 40 19 72 72
(for gas lamps)

Caving Supplies Ltd
19 London Rd
Buxton, Derbyshire
SK17 9PA
United Kingdom
tel +44 0298 71 707
fax +44 0298 72 463
(for carbide lamps)

Iain Garner Associates
PO Box 438
Harrow, Middlessex
HA2 9UT
United Kingdom
tel +44 81 868 83 53
(for kerosene lamps)

Bhawna Industries
Karnal
Haryana
132001
India
(for gas lamps)

**Changsha Tractor Accessory
Factory, Biogas office**
Hai Yan County
Zhejiang Province PRC
China
(for biogas lamps)

Coleman Gmbh
EZetilstrasse
Hungen 3
D-6303
Germany
tel +49 0640 2/89-0
fax +49 0640 2/2499
(for all lamps)

Jackwal Ltda
Av.13 de Malo, no 47 s/1102
Largo da Carioca
Rio de Janeiro
Brazil
tel +55 21 220 2449
(for biogas lamps)

Metalurgica Peneluppi Ltda
656 Rua Toledo Barbosa
Belenzinho, CEP 03061
Sao Paulo
Brazil
tel +55 291 5371
(for biogas lamps)

Patel Gas Crafters Private Ltd
20 Shree Sai Bazar
Mahatma Gandhi Road
Santacruz (W), Bombay 400 054
India
tel + 91 22 649 2501
fax +91 22 612 4424
(for biogas lamps)

Providus S.P.A
153 Corso Unione Sovietica
10134
Torino
Italy
tlx 221114 CSIND
(for gas lamps)

Vaupel Verfahrenstechnik Gmbh
Holker Feld 38
Wuppertal 2
D-5600
Germany
tel +49 0202 64 7930
(for biogas lamps)

SMALL WIND ELECTRICITY GENERATORS *

Aerodyn Wind Turbines
PO Box 2, Meadow Farm
Greenham, Wellington
TA21 0AW
United Kingdom
tel +44 82 366 61 77

Bergey Windpower Co.
2001, Priestley Avenue
Norman
OK 73069
USA
tel +1 405 364 42 12
fax +1 405 364 20 78

H Energiesystemen BV
Industrieweg 14
Swifterbant
8255 PB
The Netherlands
tel +31 3212 25 99
fax +31 3212 25 99

LMW Windenergy BV
Lijnbaanstraat 1a
9711 Groningen
The Netherlands
tel +31 50 14 52 29
fax +31 50 14 62 93

LVM Ltd
Baldock
Herts
United Kingdom
tel +44 462 89 60 95
fax +44 462 89 40 18

Marlec Engineering Co Ltd
Trevithick Rd., Corby
Northamptonshire, NN17 1XY
United Kingdom
tel +44 536 20 15 88
fax +44 536 40 02 11

NM-Electro
DK-7900
Nykobing Mors
Denmark
tel +45 9772 5200
fax +45 9772 5200

Proven Engineering Ltd
Moorfield Industrial Estate
Kilmarnock, KA2 0BA
United Kingdom
tel +44 563 43 020
fax +44 563 39 119

Rikan Engineering Ltd
PO Box 3131
Napier
New Zeland
tel +64 68 43 64 94
fax +64 68 43 64 94

SWIAB
Vettershaga 2784
S-760 10
Bergshamra
Sweden
tel +46 176 64 224
tel +46 176 64 224

Thermax Corp.
PO Box 3128
Burlington
VT 05401
USA
tel +1 802 658 1098
fax +1 802 658 1098

Vergnet SA
66 rue Hoche
Malakof
France
tel +33 1 47 46 16 16
fax +33 1 47 46 06086
tlx 632295 VERGNET F

Windsund Energy Systems
Unit 3, Industrial Estate
Spott Road, Dunbar
East Lothian, EH42 1RS
United Kingdom
tel +44 368 63 981
fax +44 368 63 981

* The wind turbines supplied by the listed manufacturers range from a few watts to a few kilowatts.

Burkardt Turbines
PO Box 1436
Dept. ASE
Ukiah
CA 95482
USA

Harris Hydroelectric
632 Swanton Road
Davenport
CA 95017
-
USA
tel +1 408 425 7652

Homestead Engineering
Highway 36
#32801, Dept. ASE
Bridgeville
CA 95526
USA

Independent Power Company
12340 Tyler Foote Road
Dept. ASE
Nevada City
CA 95959
USA

Powerhouse Paul's Turbines
PO Box 1557
Sussex
New Brunswick
EOE 1PO
Canada

* The miniature hydro-electric turbines supplied by the listed manufacturers range from a few watts to a few kilowatts.

APPENDIX B

Buyers' guide to products

During 1992/1993, lighting systems manufacturers and suppliers worldwide were contacted and asked to supply information concerning their products and services. Their replies are summarized in this appendix. The entries are grouped into types of lighting systems (electrical lighting, solar lighting and flame-based lighting). They are listed by continent where appropriate and, within this, alphabetically. Photographs have been included where they were supplied by the manufacturer. In some cases, samples of products were photographed by the authors.

It should be realized that the completeness and accuracy of the data and the quality of photographs can only be as good as has been supplied by the manufacturers. Costs are given in US dollars. The prices given should only be taken as approximate.

COMPANY:	GTE Sylvania Ltd	STATUS:	Manufacturer

CONTACT: David Hogan   TITLE: Lighting Design Engineer
ADDRESS: Otley Road, Charlestown, West Yorkshire, BD17 7SN
COUNTRY: United Kingdom
TEL: +44 274 59 59 21   FAX: +44 274 59 76 83   TLX: 51251

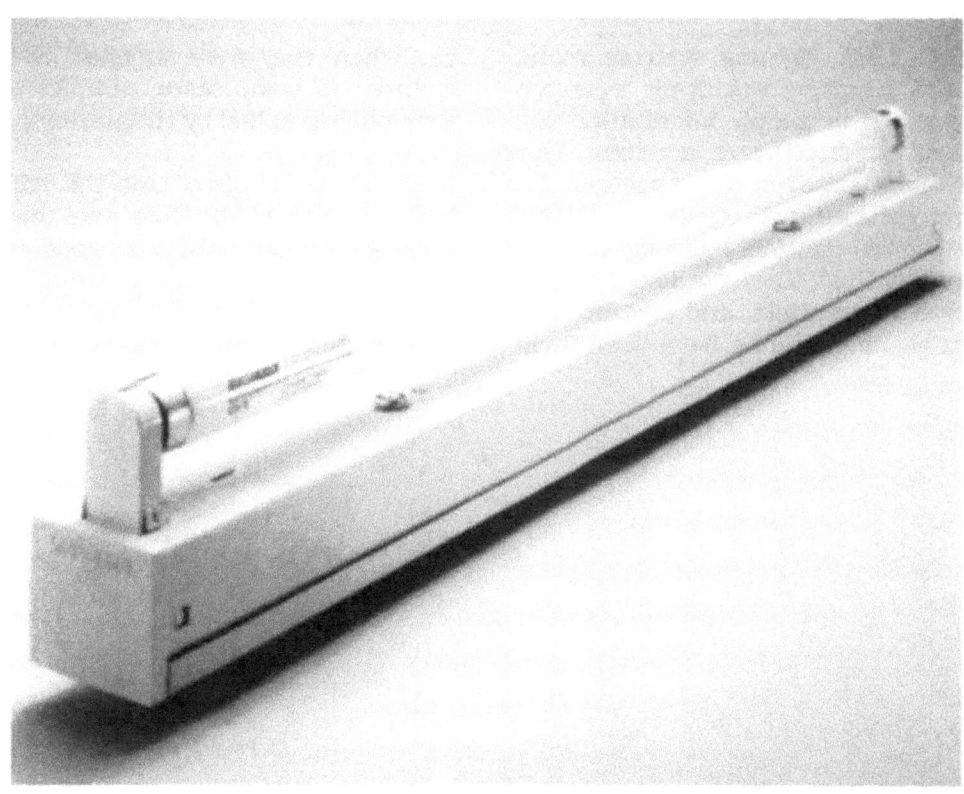

(Photo: single fluorescent tube luminaire)

TYPES OF LAMP MANUFACTURED	POWER RANGE Watt		EFFICACY RANGE Lm/W
	Y/N	MIN	MAX	MEAN
Standard tungsten lamps	Yes	15	1000	13
Tungsten halogen lamps	Yes	100	2000	20
Fluorescent lamps	Yes	5	215	55
CFL with electronic ballast*	Yes	7	20	60
CFL with magnetic ballast	Yes	-	-	-
CFL separate tube & ballast	Yes	5	36	80
Sodium lamps	Yes	35	1000	115
Mercury/Metal halide lamps	Yes	50	1000	70
Other lamps	Yes	-	-	-

* CFL means Compact Fluorescent Lamps

OTHER PRODUCTS: Luminaires for tubular and compact fluorescent lamps
Luminaires for halogen and HID lamps, floodlights, etc.

COMPANY: **Moorlite Electrical Ltd**   STATUS: Manufacturer

CONTACT: Tracy Cuddy   TITLE: Lighting Engineer
ADDRESS: Burlington Street, Ashton-Under-Lyne, Lancashire, OL7 0AX
COUNTRY: United Kingdom
TEL: +44 61 330 68 11   FAX: +44 61 330 28 15   TLX: 668284

PRODUCTS:

o Luminaires for fluorescent tubes
  of different construction, surface
  mounted, recesseed or suspended
o Luminaires for CFLs
o Luminaires for special environ-
  ments (hazardous, prison, etc.)
o Fluorescent control gear
  including dimmable types

SERVICES:
o Free computer-assisted design
  service available for designers

(Photo: Luminaires for CFLs)

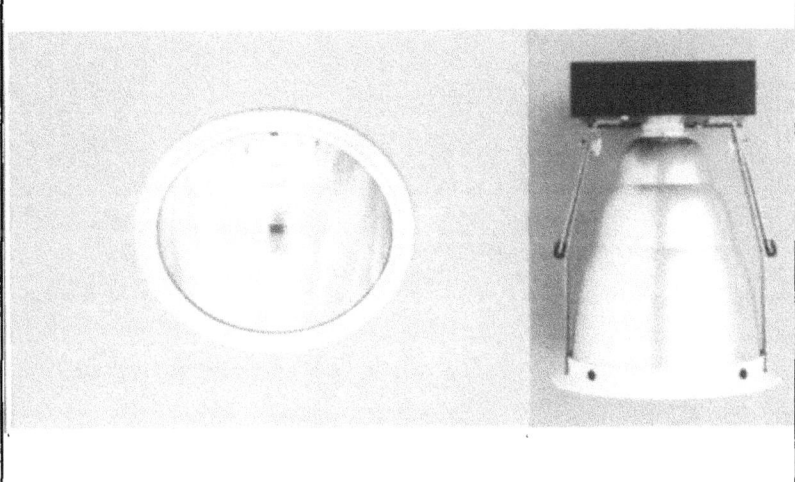

COMPANY: **Progressive Technology in Lighting**   STATUS: Manufacturer

CONTACT: Kevin Youngquifp   TITLE: Director
ADDRESS: 581 Ottawa Avenue, Holland, MI 49423
COUNTRY: USA
TEL: +1 616 396 25 56   FAX: +1 616 396 06 86   TLX:

PRODUCTS:

o Adapters for compact fluorescent
  lamps CFL from 5 to 13 W to screw into
  any standard incandescent sockets
o Reflectors and diffusers adapted
  to different CFL sizes
o Convertor sets to use CFL on any
  lighting fixtures using a cord
  and wall plug (e.g. table lamps)
o CFL kits for exit signs
o CFL of different appearance and
  sizes

(Photo: Various CFL fittings)

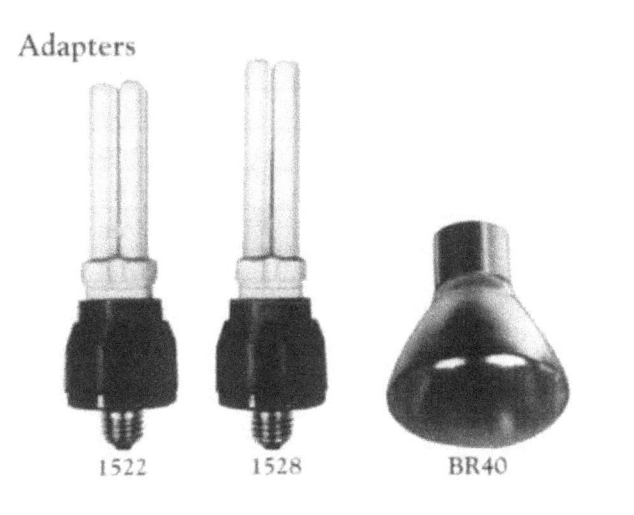

Adapters

1522   1528   BR40

ELECTRICAL LIGHTING

COMPANY:	Osram Gmbh	STATUS:	Manufacturer

CONTACT: **B. Stellmacher** TITLE: **Sales Manager**
ADDRESS: **81543, Munich**
COUNTRY: **Germany**
TEL: **+49 896 21 31** FAX: **+49 896 21 32 020** TLX:

(Photo: various types of compact fluorescent lamps)

TYPES OF LAMP MANUFACTURED	Y/N	POWER RANGE Watt		EFFICACY RANGE Lm/W	
		MIN	MAX	MIN	MAX
Standard tungsten lamps	Yes	25	500	7	11
Tungsten halogen lamps	Yes	20	1500	12	25
Fluorescent lamps	Yes	4	125	32	80
CFL with electronic ballast*	Yes	7	23	57	60
CFL with magnetic ballast	No	-	-	-	-
CFL separate tube & ballast	Yes	5	55	57	75
Sodium lamps	Yes	18	1000	70	180
Mercury/Metal halide lamps	Yes	50	1000	16	60
Other lamps	Yes	-	-	-	-

* CFL means Compact Fluorescent Lamps

OTHER PRODUCTS: Special lamps for traffic signals, vehicles, projection, etc...
Electronic control gear for fluorescent tubes, CFLs, HID, etc.

ELECTRICAL LIGHTING

COMPANY:	Philips Lighting	STATUS:	Manufacturer

CONTACT: E. Schoenmakers   TITLE: General Manager
ADDRESS: BV Building ED1, PO Box 80020, 5600 JM, Eindhoven
COUNTRY: The Netherlands
TEL: +31 40 75 70 65   FAX: +31 40 75 54 53   TLX: 35000 PHTC

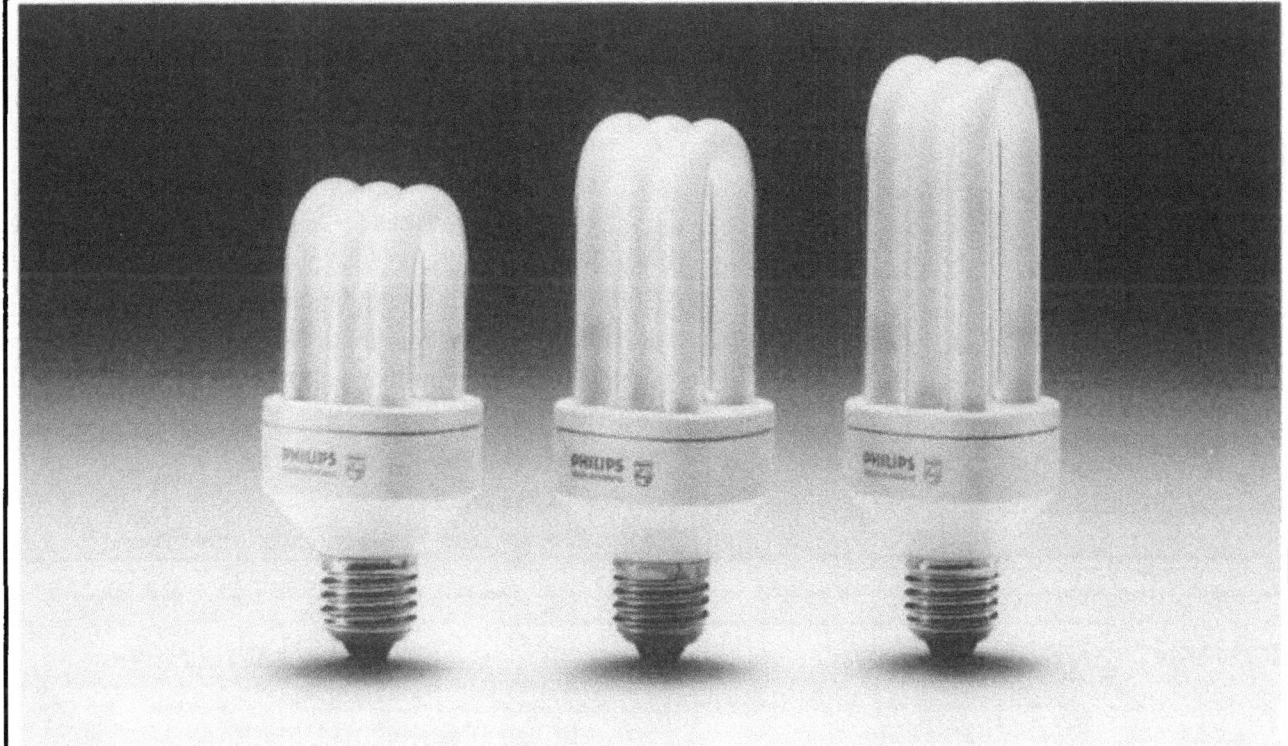

(Photo: Compact fluorescent lamps with folded tubes)

TYPES OF LAMP MANUFACTURED	Y/N	POWER RANGE Watt		EFFICACY RANGE Lm/W	
		MIN	MAX	MIN	MAX
Standard tungsten lamps	Yes	0.75	1000	10	17
Tungsten halogen lamps	Yes	2	2000	15	35
Fluorescent lamps	Yes	4	125	40	100
CFL with electronic ballast*	Yes	9	23	45	60
CFL with magnetic ballast	Yes	9	25	45	50
CFL separate tube & ballast	Yes	5	26	50	65
Sodium lamps	Yes	18	1000	110	180
Mercury/Metal halide lamps	Yes	50	2000	50	80
Other lamps	Yes	-	-	-	-

* CFL means Compact Fluorescent Lamps

OTHER PRODUCTS: Luminaires, amenity and security lighting, street lighting
Electronic control gear for any kind of lamps
Dry cells and rechargeable batteries for consumer and industrial
applications. Special lamps such as projection lamps, car lights, etc.

COMPANY: **Silverlight** STATUS: Manufacturer

CONTACT: **Anthony Reiss** TITLE: **Manager**
ADDRESS: **Maidstone Road, Matfield, Tonbridge, Kent, TN12 7JN**
COUNTRY: **United Kingdom**
TEL: **+44 892 72 22 02** FAX: **+44 872 72 35 07** TLX: 95539 BERK G

PRODUCTS:

Range of energy-efficient luminaires
with specular reflectors:

- o recessed luminaires
- o batten luminaires
- o surface mounted luminaires
- o sealed luminaires

SERVICES:

o Complete design including: number, sizes
and costs of luminaires, scheme-drawing
detailing luminaire positions and
switching circuitry, installation
specification and tender package for
contractors to price.

(Photo: Luminaire with high specular reflectors)

COMPANY: **Verre & Quartz** STATUS: Manufacturer

CONTACT: **Jean-Louis Diot** TITLE: **Marketing Director**
ADDRESS: **24 route d'Aulnay, 93147 Bondy Cedex**
COUNTRY: **France**
TEL: **+33 1 48 48 14 22** FAX: **+33 1 48 48 68 18** TLX: 233593F

PRODUCTS:

o Special lighting systems for health care:
- general medical examination
- minor surgery
- gynaecology
- dental applications
- etc.

Lighting appliances equipped with:
- 12/24 V DC halogen lamps
- 220 V AC compact fluorescent lamps

(Photo: Medical examination lamp)

| COMPANY | BP Solar International Ltd | STATUS: | Manufacturer |

CONTACT: Rod Scott   TITLE: Technology Manager
ADDRESS: Solar House, Bridge Street, Leatherhead, Surrey KT22 8BZ
COUNTRY: United Kingdom
TEL: +44 372 37 78 99   FAX: +44 372 37 77 50   TLX: 263220 BPSIL G

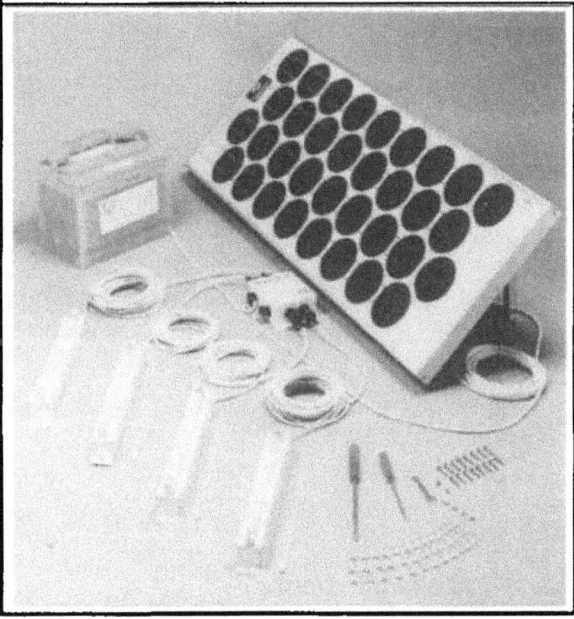

TYPE OF SYSTEMS:	Portable lamp	Lighting kit
Models	SL 48	LK 4
Operating voltage (V)	6	12
Number of lamps	1	4
Watt per lamp (W)	8	8
Lumens per lamp (lm)	400	400
Panel peak power (Wp)	10	50
Battery capacity (Ah)	4	100
Cost (US$)	350	750

OTHER PRODUCTS: Lighting kits LK5, LK9, street lighting system with low pressure sodium or fluo. lamps, PV modules, refrigerators, pumps, power packs.
SERVICES PROVIDED: Site evaluation, system selection, installation and maintenance
REPRESENTED IN:
Australia, Indonesia, Thailand, Kenya, South Africa, etc.
(Photo: Lighting kit LK 4)

| COMPANY: | Ecosolaire / Serelio | STATUS: | Manuf/Distributor |

CONTACT: Jean-Claude Bernard   TITLE: Managing Director
ADDRESS: 19 rue Pavée, 75004 Paris
COUNTRY: France
TEL: +33 1 48 87 43 60   FAX: +33 1 48 87 86 27   TLX: 214235F

TYPE OF SYSTEMS:	Portable lamps	
Models	TS 15	ME
Operating voltage (V)	12	6/12
Number of lamps	1	2
Watt per lamp (W)	5	6
Lumens per lamp (lm)	250	300
Panel peak power (Wp)	5	2
Battery capacity (Ah)	-	4
Cost (US$)	150	110

OTHER PRODUCTS: Customized lightings kits, solar pumps, refrigerators, public address and ventilation systems, charge regulators, solar ni-cad chargers
SERVICES PROVIDED:
Technical advice and demonstration
REPRESENTED IN: France

(Photo: Portable lantern TS 15)

SOLAR LIGHTING

COMPANY	Hiltec Solar Ltd	STATUS:	Manuf/Distributor

CONTACT: **Ben Hill** TITLE: **Director**
ADDRESS: **25 Low Friar Street, Newcastle upon Tyne, NE1 5UE**
COUNTRY: **United Kingdom**
TEL: **+44 91 232 88 18** FAX: **+44 91 261 57 46** TLX:

TYPE OF SYSTEMS:

o solar panels and lanterns
o lanterns with PV panels incorporated
o solar torches
o universal solar ni-cad chargers (e.g. Unisol)
o solar short-wave radios
o PV modules

SERVICES PROVIDED:
o technical advice and demonstration

(Photo: Solar ni-cad charger 'Unisol')

COMPANY:	Real Goods	STATUS:	Manuf/Distributor

CONTACT: **John Schaeffer** TITLE: **Director**
ADDRESS: **966 Mazzoni Street, CA 95482-3471**
COUNTRY: **USA**
TEL: **+1 800 762 7325** FAX: **+1 707 468 0301** TLX:

TYPE OF SYSTEMS

o solar flashlights and torches
  of different construction and types
o portable solar lanterns
o solar Ni-cad chargers
o PV modules
o solar radios
o other 12 V DC components such as
  fridges, small inverters

SERVICES PROVIDED:
o yearly publication of an Alternative Energy
  Sourcebook
o system selection and design

(Photo: Solar flashlight)

SOLAR LIGHTING

COMPANY: IBC

STATUS: Manuf/Distributor

CONTACT: Udo Möhrstedt   TITLE: Dipl. Phys.
ADDRESS: Am Hochgericht 15, PO Box 1107, D-8623 Staffelstein
COUNTRY: Germany
TEL: +49 95 73 30 66   FAX: +49 95 73 31 264   TLX:

TYPE OF SYSTEMS:	Lighting kits			Street lighting	
Models	4050	4055	4060	4080	4081
Operating voltage (V)	12	12	24	12	12
Number of lamps	2	2	3	1	1
Watt per lamp (W)	7	7	11	18	35
Lumens per lamp (lm)	-	-	-	-	-
Panel peak power (Wp)	11	25	100	100	150
Battery capacity (Ah)	50	50	130	180	180
Cost (US$)	330	400	1000	2600	2950

OTHER PRODUCTS: Solar batteries, charge regulators, inverters, complete PV systems for home, clinic, village energy supply, PV modules.
SERVICES PROVIDED: System selection, installation and maintenance
REPRESENTED IN: Tunisia, Morocco, Egypt, Nigeria, etc.

(Photo: IBC Street-lighting system)

COMPANY: Kyocera Corporation

STATUS: Manufacturer

CONTACT: Tetsuo Yasutake   TITLE: Manager Solar Energy Div.
ADDRESS: Karasuma office, 680 Karasuma Bukkoji-sagaru, Shimogyo-ku
COUNTRY: Kyoto 600, Japan
TEL: +81 75 344 8241   FAX: +81 75 344 8240   TLX:

TYPE OF SYSTEMS:	Portable lantern		Lighting kits		
Models	LS623	LS624	KLK34	KLK45	KLK96
Operating voltage (V)	6	6	12	12	12
Number of lamps	2	2	2	2	4
Watt per lamp (W)	6	6	20	20	20
Lumens per lamp (lm)	200	200	750	750	750
Panel peak power (Wp)	2.65	3.6	34	45	96
Battery capacity (Ah)	4	4	50	70	200
Cost (US$)	75	90	360	450	1100

OTHER PRODUCTS: Large selection of polycristalline PV modules, customized lighting kits, etc.

SERVICES PROVIDED: Site evaluation, system selection, installation and maintenance
REPRESENTED IN: Kenya, Indonesia, Singapore, Philipines Pakistan, Malaysia, etc.
(Photo: Solar lantern LS624)

SOLAR LIGHTING

COMPANY	Neste (Naps)	STATUS:	Manufacturer

CONTACT: Asko Rasinkoski   TITLE: R&D Manager
ADDRESS: Rälssitie 7, Vantaa, SF-01510
COUNTRY: Finland
TEL: +358 0 450 57 53   FAX: +358 0 826 301   TLX: 124641 NESTE

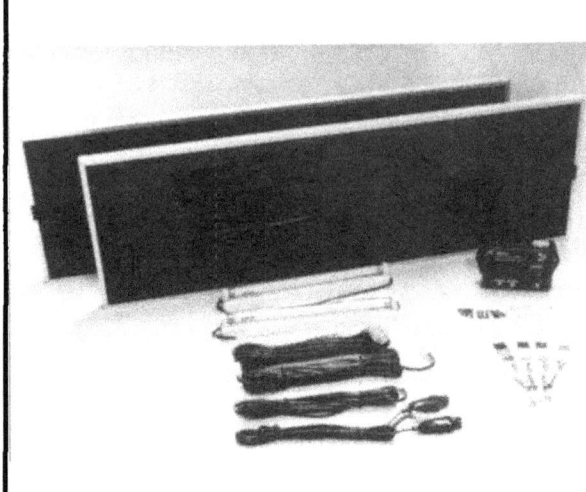

TYPE OF SYSTEMS:	Portable lamp	Lighting kit
Models	ML	Minikit
Operating voltage (V)	12	12
Number of lamps	1	2
Watt per lamp (W)	3/6.5	8
Lumens per lamp (lm)	160/320	400
Panel peak power (Wp)	7.5	18/22
Battery capacity (Ah)	7	-
Cost (US$)	150	260

OTHER PRODUCTS: Lighting kits NLK1, 2 and 3, street-lighting system with low pressure sodium or fluo. lamps PV modules, refrigerators, pumps, power packs, etc.
SERVICES PROVIDED: Site evaluation, system selection, installation and maintenance
REPRESENTED IN: in most developed and developing countries
(Photo: Naps 'Minikit' lighting system)

COMPANY:	Photowatt International SA	STATUS:	Manufacturer

CONTACT: Robert de Franclieu   TITLE: Commercial Director
ADDRESS: 33 Rue Saint-Honoré / ZA Champfleuri, 38300 Bourgoin-Jallieu
COUNTRY: France
TEL: +33 74 93 80 20   FAX: +33 74 93 80 40   TLX:

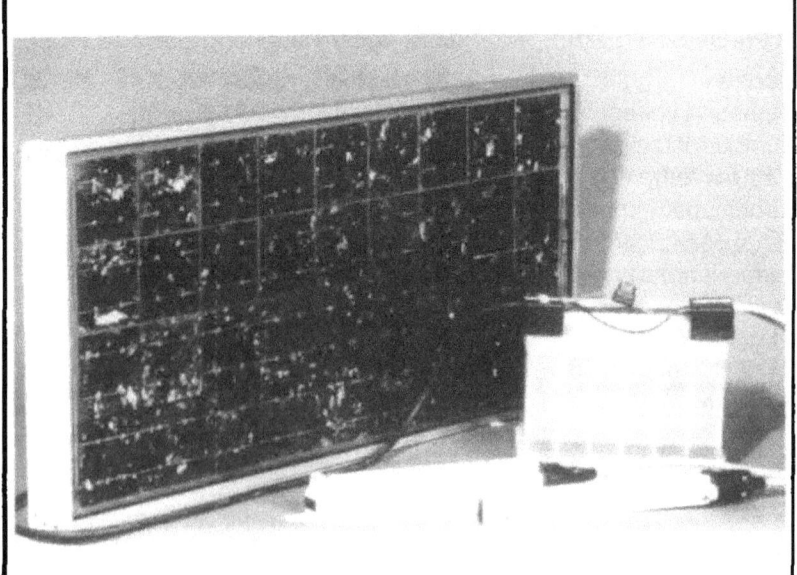

TYPE OF SYSTEMS:	Lighting kits
Models	Genedom
Operating voltage (V)	12
Number of lamps	2
Watt per lamp (W)	6
Lumens per lamp (lm)	300
Panel peak power (Wp)	35/50
Battery capacity (Ah)	-
Cost (US$)	450

OTHER PRODUCTS:
Various lighting kits, pumps, refrigerators phone booths, charge regulators, etc.
SERVICES PROVIDED:
System sizing and design
REPRESENTED IN: West African countries
(Photo: Lighting kit 'Genedom')

SOLAR LIGHTING

| COMPANY: | Solar Products International Ltd | STATUS: | Distributor |

CONTACT: Iain Garner   TITLE: Director
ADDRESS: PO Box 438, Harrow, Middlesex, HA2 9UT
COUNTRY: United Kingdom
TEL: +44 81 868 83 53   FAX: +44 81 429 42 70   TLX:

TYPE OF SYSTEMS:	Portable lamps		lighting kits		
Models	120GL	SUNX2	MINI	MIDI	MAXI
Operating voltage (V)	4	6	12	12	12
Number of lamps	1	2	2	4	8
Watt per lamp (W)	6	6	9	9	9
Lumens per lamp (lm)	300	300	650	650	650
Panel peak power (Wp)	1.3	3.6	10	48	96
Battery capacity (Ah)	4.1	4	50	105	210
Cost (US$)	100	130	440	800	1460

OTHER PRODUCTS: Solar flashlights, pumping systems solar radios, refrigeration and street lighting

SERVICES PROVIDED: System sizing, component selection site evaluation and training
REPRESENTED IN: Kenya, Botswana, Mali, South Africa, Namibia, Caribbean, Southern Pacific, Middle East and Asia
(Photo: Solar lantern SUNX2)

| COMPANY: | Solec International Inc. | STATUS: | Manufacturer |

CONTACT: Jawid A. Shahryar   TITLE: Sales & Marketing Manager
ADDRESS: 12533 Chadron Ave., Hawthorne, CA 90250
COUNTRY: USA
TEL: +1 301 970 00 65   FAX: +1 301 970 10 65   TLX: 910 325 6215

TYPE OF SYSTEMS:	Security and area lighting systems				
Models	WFL25	70PM	90PM	180PM	270P
Operating voltage (V)	12	12	24	24	24
Number of lamps	1	1	1	1	1
Watt per lamp (W)	9	18	35	55	50
Lumens per lamp (lm)	600	1800	4800	8000	3800
Panel peak power (Wp)	25	70	90	180	270
Battery capacity (Ah)	100	200	200	400	200
Cost (US$)					

OTHER PRODUCTS: Solar PV modules, battery chargers, solar-powered tool-kits, military man-pack solar array

SERVICES PROVIDED: System selection, sizing, installation and maintenance
REPRESENTED IN:

(Photo: Street-lighting system SOLEC)

SOLAR LIGHTING

COMPANY: **Soltech Ltd**   STATUS: | Manufacturer

CONTACT: **Joris De Vos**   TITLE: **Director**
ADDRESS: **Kapeldreef 75, B-3550 Leuven**
COUNTRY: **Belgium**
TEL: **+32 16 27 04 42**   FAX: **+32 16 27 04 43**   TLX: **26152**

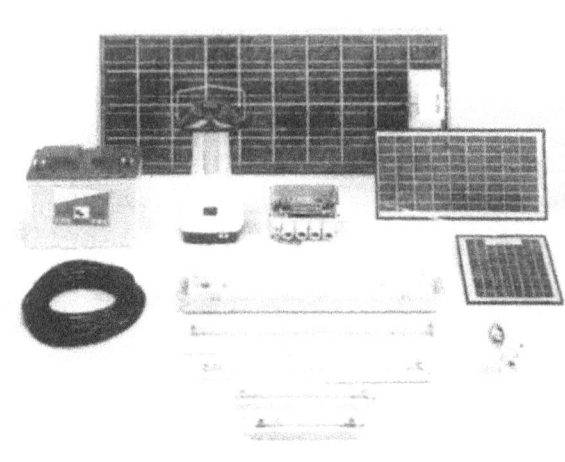

TYPE OF SYSTEMS:	Portable la	Lighting kits		
Models	P9	4L8	3L13	2L18
Operating voltage (V)	4	12	12	12
Number of lamps	1	4	3	2
Watt per lamp (W)	6	8	13	18
Lumens per lamp (lm)	300	400	800	1100
Panel peak power (Wp)	1.26	40	40	40
Battery capacity (Ah)	4.1	100	100	100
Cost (US$)	280	850	850	850

OTHER PRODUCTS: Street-lighting systems charge regulators and inverters

SERVICES PROVIDED: Site evaluation and system selection, installation and maintenance
REPRESENTED IN:

(Photo: Various Soltech lighting systems)

COMPANY: **Vergnet S.A.**   STATUS: | Manuf. / Supplier

CONTACT: **Jérôme Billeret**   TITLE: **Solar Department Manager**
ADDRESS: **6 rue Henri Dunant, 45140 Ingre**
COUNTRY: **France**
TEL: **+33 38 43 36 52**   FAX: **+33 38 88 30 50**   TLX: **780980F**

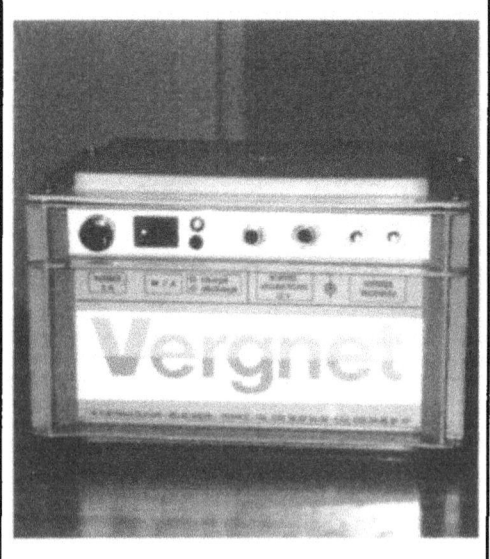

TYPE OF SYSTEMS:	Lighting kits			Power packs	
Models	KE1	KE2	KE3	BE7	BE14
Operating voltage (V)	12	12	12	12	12
Number of lamps	3	5	10	1	2
Watt per lamp (W)	8	8/13	8/13	8	8
Lumens per lamp (lm)	300	-	-	300	300
Panel peak power (Wp)	48	96	192	11	22
Battery capacity (Ah)	105	105	210	7	14
Cost (US$)				230	400

OTHER PRODUCTS: Solar pumps, refrigerators, public address and ventilation systems, wind generators, power packs with extra capacity and 220V AC
SERVICES PROVIDED: System sizing, installation and maintenance by local representatives
REPRESENTED IN: almost all french speaking countries in Africa
(Photo: Power pack BE7)

COMPANY:	Caving Supplies Ltd	STATUS:	Manuf./Distributor

CONTACT: Phil Brown TITLE: Manager
ADDRESS: 19 London Road, Buxton, Derbyshire, SK17 9PA
COUNTRY: United Kingdom
TEL: +44 298 71 707 FAX: +44 298 72 463 TLX:

PRODUCTS:	PRICE (US$)
ACETYLENE:	
o Carbide caplamps and headlights	33-70
o Carbide generators	42
o Carbide in 1, 3, 25 and 100kg tins	3/kg
o Lamp and generator spare parts	
DC ELECTRIC:	
o Electric mining lamps and batteries	30-100
o Battery chargers and adaptors	
o Dry battery electric lamps	
KEROSENE:	
o Brass Miners oil lamps (Davy lamp)	70

(photo: Carbide lamp PL85-27)

COMPANY:	Iain Garner Associates	STATUS:	Distributor

CONTACT: Iain Garner TITLE: Director
ADDRESS: PO Box 438 Harrow, Middlesex, HA2 9UT
COUNTRY: United Kingdom
TEL: +44 81 868 83 53 FAX: +44 81 868 83 53 TLX:

PRICE (US$)	PRODUCTS:
8	o Hurricane lanterns (paraffin wick)
60	o Pressure lamps (paraffin mantle) US and UK models:
	- height: 34 cm
	- width: 18 cm
	- weight: 2.3 kg (empty)
	- vitreous enamelled hood
	- heat-resisting glass globe
	- polished brass tank
	- supplied with funnel with filter, spirit bottle and instructions.

(Photo: Kerosene pressure lamp)

APPENDIX C

Calculation of the daylight factor

Methodology

The extent to which direct or indirect daylight contributes to the lighting of an interior is evaluated by the mean of the 'daylight factor,' (ADF) which can be averaged in an empty interior using the following approximate expression:

$$ADF = (T \times A_w \times d) / A_{is} \times (1 - R^2)$$

where:

ADF = average daylight factor (%);

T = transmittance of glazing material (decimal), (T = 1 when there is no glazing) (see Table 5.2 'Transmittance values of several diffuser materials for visible light');

A_w = net area of glazing (m²);

d = angle (degrees) subtended by sky visible from the centre of a window (figure below);

A_{is} = total area of ceiling, floors and walls, including windows (m²);

R = area weighted average reflectance of all indoor surfaces, including windows (see Table 5.4 'Average reflectance values for visible light').

The above formula applies for both windows and skylight.

Some rooms may appear to be fairly evenly day-lit. These include many roof-lit and many shallow side-lit rooms. In a deep side-lit room, however, the lighting in the depth of the interior may look very dull when compared with the lighting just inside the window. This

---

Figure C.1 Determination of the angle subtended by sky visible from the centre of a window

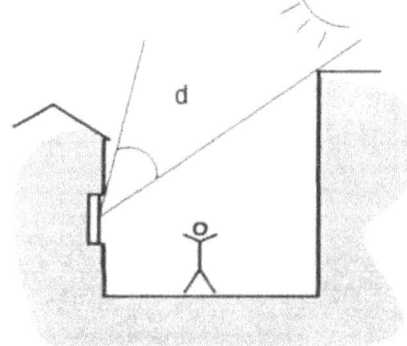

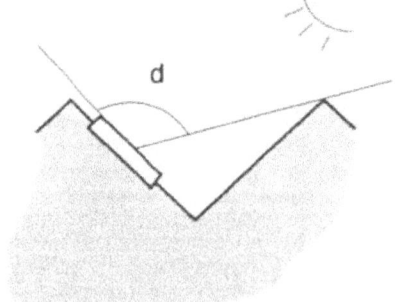

a) d is the angle subtended, in a vertical plane normal to the window, by the sky visible from the centre of the window

b) d is the angle subtended, in a vertical plane normal to the roofwindow, by the sky visible from the centre of the roofwindow

is likely to occur when the depth of the room from window to back wall is greater than the limiting depth calculated from the expression:

$$D = (2 \times W \times H) / [(H + W) \times (1 - R_b)]$$

where:

D - limiting depth (m);

W - width of room from side to side parallel to window (m);

H - height of top of the window above floor (m);

R_b - area weighted average reflectance of surfaces in the half of the room remote from the window (%).

For rooms lit from two opposite sides, the limiting depth may be doubled.

The above calculations refer to an average daylight factor and should be enough for most cases in rural lighting. For those special

---

Example of ADF calculation

Lighting is needed for a room A in a health centre. The room has the following dimensions: width 3m, depth 3m and height between floor and ceiling is 2.8m. The window is: 2m wide x 1.2m high (or two small windows with the same total area) and is made of clear glass. The base of the window is at 1.2m from the floor. The walls and ceiling have been freshly painted in white. The room is situated in front of a building of 4m high and situated 4m away. The same room is referred to in Chapter 9, Table 9.2.

1. The situation is depicted in Figure C.2 on the following page

2. As seen in the drawing, the angle d is 45°

3. Area of the window is 2 x 1.2 = 2.4m²

4. Transmittance of the window is : 0.9 (see Chapter 5,Table 5.2)

5. Area of all indoor areas (A) and area weighted average reflectance (R) (see Chapter 5, Table 5.4)

	Area	Reflectance
Ceiling	9.0	70% or 0.7
Wall	31.2	50% or 0.5
Floor	9.0	20% or 0.2
Window	2.4	30% or 0.3
	$A_{ts} = 51.6$	

$$R = \frac{9 \times 0.7 + 31.2 \times 0.5 + 9 \times 0.2 + 2.4 \times 0.3}{51.6}$$

$$R = 0.47$$

6. ADF = 0.9 x 2.4 x 45 / (51.6 x (1 - 0.47²)) = 2.42 %

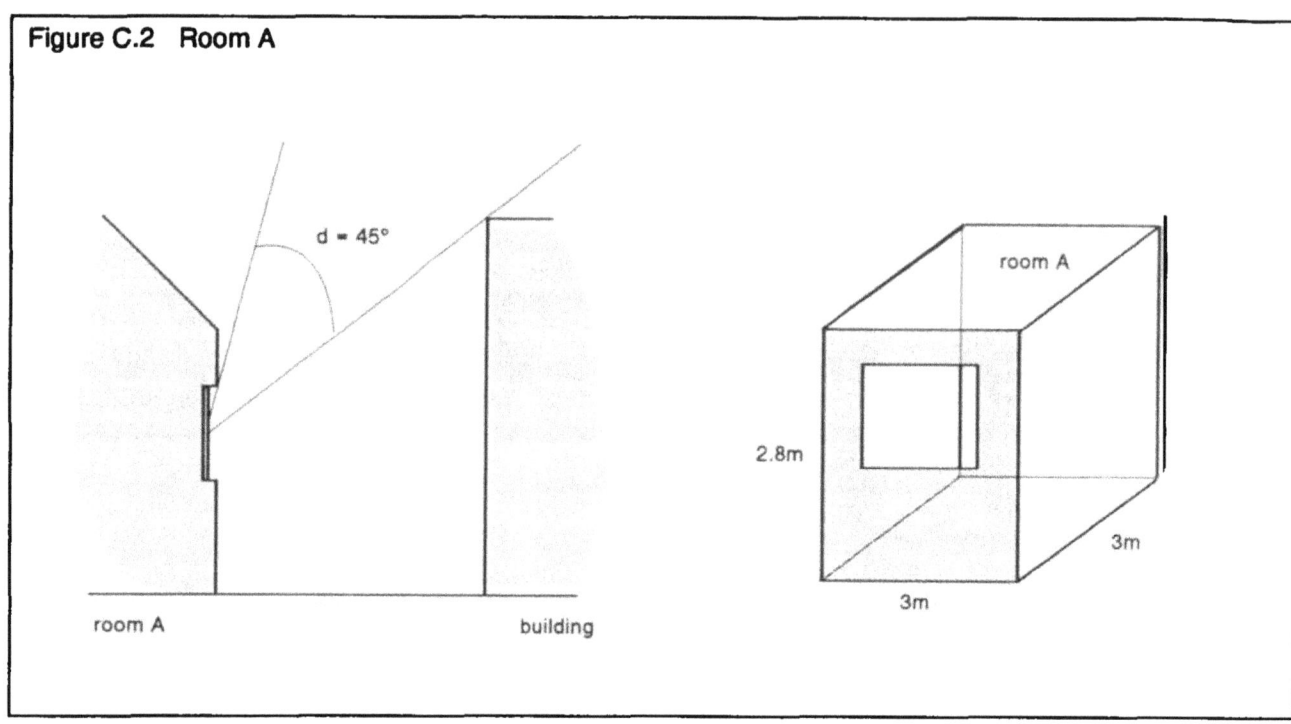

Figure C.2   Room A

designs where point-by-point daylight factor calculations are necessary, several methods are available. One of these methods can be found in the *Book Code for Interior Lighting* (see Bibliography).

Comment on the example calculation

As the ADF is below 5%, it is likely that a lighting system will be required during the day (see Table C.1, 'Influence of average daylight factor values on electric lighting'). Nevertheless, if there is the possibility of constructing this window on another side of the building with no building in front of it, then the angle d will be equal to 90° which doubles the ADF. In this case, the average daylight factor would be close to 5% and a lighting system during the day is not necessary.

Table C.1   Influence of average daylight factor values on electric lighting

ADF = Average daylight factor

ADF < 2%

Electrical lighting in use throughout the day

2% < ADF < 5%

Localised or local electric lighting with daylight for general lighting

ADF > 5%

Minimal or no use of electric lighting during the day

APPENDIX D

Calculation of the utilization factor

Methodology

The Utilization Factor (UF) of a lighting installation is the ratio of the total lumens received by the reference plane to the total lumens emitted by the lamps. It depends on the luminaire characteristics, the reflectance of the surfaces and room proportions. To calculate the utilization factor, it is necessary to go through the following steps below.

Note: Before starting this calculation, appropriate manufacturers data on luminaires must be available. They usually can be found in technical catalogue from manufacturer. If no data are available, use Table 9.1 in Chapter 9. This table gives UF values for the most common cases. For other cases extrapolation might be necessary.

1. Determination of the Room Index (RI)

The room index is a measure of the proportions of the room. It is given for rectangular rooms by the following expression:

$$RI = (L \times W) / ([L + W] \times H)$$

where:

L = length of the room (m);

W = width of the room (m);

H = height of luminaire above horizontal reference plane (m).

If a room has a complex shape (e.g. L-shaped or U-shaped), it should be divided into several rectangular sections which can be dealt with separately.

2. Determination of the Effective Reflectances

In a room, the space above the luminaire plane is called the ceiling cavity, and the space below the horizontal working plane is called the floor cavity (See Figure D.1 'Deter-

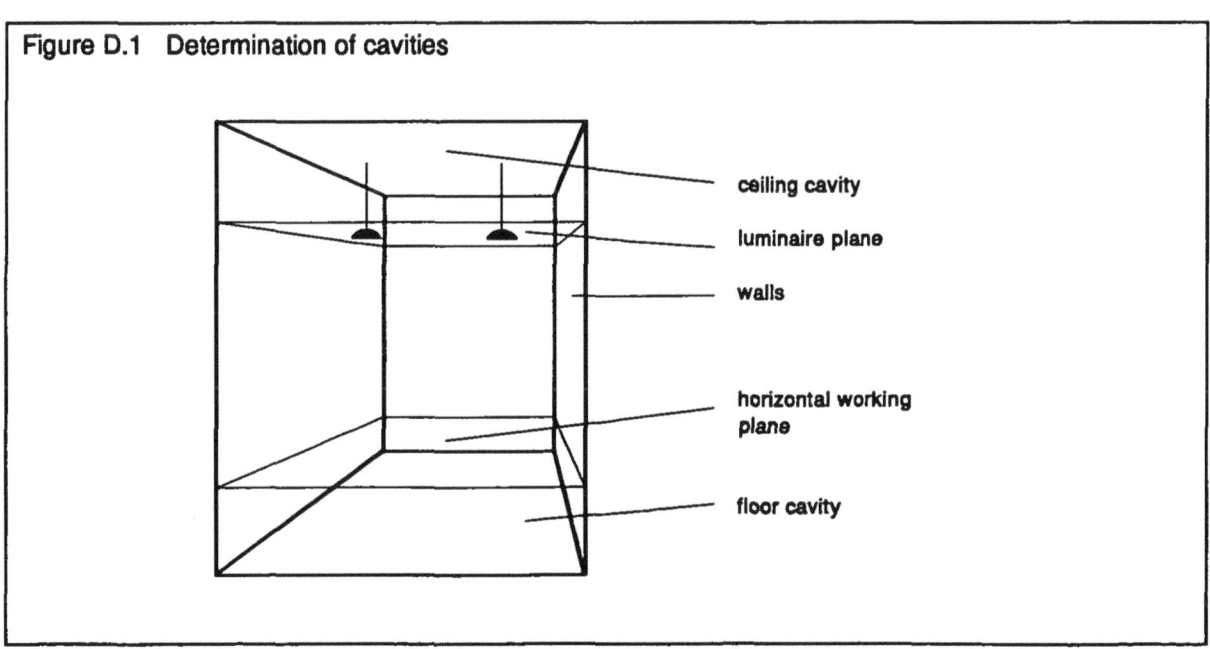

Figure D.1   Determination of cavities

- ceiling cavity
- luminaire plane
- walls
- horizontal working plane
- floor cavity

mination of cavities'). For these cavities respectively, the cavity index CI(C) and CI(F), which is similar in concept to the room index, are given in a rectangular room by:

$$CI(C) \text{ or } CI(F) = (L \times W) / ([L + W] \times h)$$
$$= RI \times H / h$$

where h is the depth of the cavity (e.g. for a CI[F], h is the distance between the floor and the working plane).

The average reflectance RA(X) of a series of surfaces can be determined from the following formula:

$$RA(X) = ([sum \text{ of } 1 \text{ to } n] Rk \times Sk) /$$
$$([sum \text{ } 1 \text{ to } n] \times Sk)$$

where:

Sk = respective areas of the surfaces 1 to n of the series (m²);

Rk = respective reflectances of the surfaces 1 to n of the series.

The effective reflectance RE(X) of a cavity is determined from the following formula:

$$RE(X) = [CI(X) \times RA(X)] / [CI(X) + 2(1 - RA(X)]$$

where:

RA(X) = average reflectance of the cavity X;

CI(X) = cavity index of the cavity X.

3. Determination of the Utilization Factor (UF)

It is determined from the manufacturer's data for the chosen luminaire, using the room index and the effective reflectance of wall (W), ceiling (C) and floor (F) as determined above. An example of such a table for a given type of luminaire (in this case, a luminaire for one fluorecent tube with reflector) is given in Figure D.2.

For example, if RE(C) is 0.50, RE(W) 50, RE(F) 20 and RI 2, then UF = 73%.

Figure D.2   Example table of utilization factors for a luminaire (courtesy of Philips Lighting)

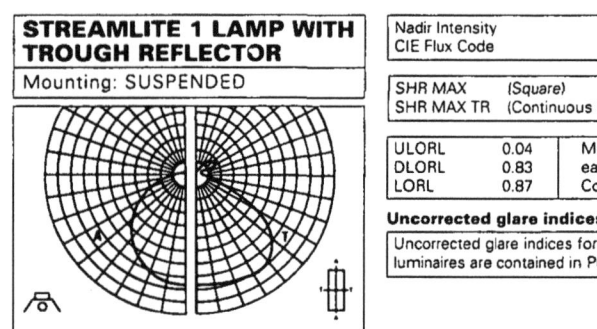

STREAMLITE 1 LAMP WITH TROUGH REFLECTOR
Mounting: SUSPENDED

Nadir Intensity	241cd/1000lm
CIE Flux Code	43 79 96 95 87

SHR MAX	(Square)	2.01 (2.00 NOM)
SHR MAX TR	(Continuous Rows)	2.36

ULORL	0.04	Multiply by each Service Correction Factor
DLORL	0.83	
LORL	0.87	

Uncorrected glare indices

Uncorrected glare indices for all Philips indoor luminaires are contained in Publication PL4412.

Service Correction Factors

	36W	58W	70W	100W
Length Factor	1.00	1.00	1.00	0.99
Colours 80 Factor	1.00	1.00	1.00	1.00
38mm Factor	0.98	0.98	0.98	1.00
Ballast Lumen Factor	1.00	1.00	1.00	1.00
Operating Factor K	0.85	0.85	0.85	0.85
Emergency Ballast Lumen Factor	0.27	0.15	0.13	0.07

Utilization Factors UF(F) for SHR NOM

Room Reflectances			Room Index								
C	W	F	0.75	1.00	1.25	1.50	2.00	2.50	3.00	4.00	5.00
70	50	20	55	59	65	70	76	80	83	86	89
	30		49	53	59	64	71	75	78	83	86
	10		44	48	54	59	66	71	75	80	83
50	50	20	53	57	62	67	73	76	79	82	84
	30		48	51	57	62	68	72	75	79	82
	10		44	47	53	58	65	69	72	77	80
30	50	20	51	55	60	64	69	73	75	78	80
	30		47	50	55	60	66	70	72	76	78
	10		43	46	52	56	63	67	70	74	77
0	0	0	41	44	49	53	59	63	66	69	72

Multiply by each Service Correction Factor

Test No. C129	Date 81.04.02
Measured in accordance with BS 5225	Calculations based on CIBSE TM5 and TR10

Single future costs or benefits

Table E.1 is used to find the discount factor for a single cost or benefit at some point in the future. The future cost should be multiplied by the appropriate factor to give its present worth. A description and examples of the use of this and the following table for the calculation of present worth are given in section 10.3.

Table E.1   Discount factors, Pr, for a single future payment

Years	Discount rate (d)					
	0.04	0.06	0.08	0.10	0.12	0.14
1	0.96	0.94	0.93	0.91	0.89	0.88
2	0.92	0.89	0.86	0.83	0.80	0.77
3	0.89	0.84	0.79	0.75	0.71	0.67
4	0.85	0.79	0.74	0.68	0.64	0.59
5	0.82	0.75	0.68	0.62	0.57	0.52
6	0.79	0.70	0.63	0.56	0.51	0.46
7	0.76	0.67	0.58	0.51	0.45	0.40
8	0.73	0.63	0.54	0.47	0.40	0.35
9	0.70	0.59	0.50	0.42	0.36	0.31
10	0.68	0.56	0.46	0.39	0.32	0.27
12	0.62	0.50	0.40	0.32	0.26	0.21
15	0.56	0.42	0.32	0.24	0.18	0.14
20	0.46	0.31	0.21	0.15	0.10	0.07

The formula used to calculate the single payment factors, **Pr**, is:

$$Pr = [\,(1 + i)\,/\,(1 + d)\,]^{n}$$

i - inflation rate over and above the general rate (i is assumed to be zero)

d - discount rate

n - number of years in the future

Annually recurring costs or benefits

Table E.2 is used to find the discount factor for an annually recurring cost or benefit. The annual figure should be multiplied by the appropriate factor to give the total cumulative present worth of those costs over the life-cycle.

Table E.2 Discount factors, Pa, for annually recurring costs

Years	Discount rate (d)					
	0.04	0.06	0.08	0.10	0.12	0.14
5	4.45	4.21	3.99	3.79	3.60	3.43
6	5.24	4.92	4.62	4.36	4.11	3.89
7	6.00	5.58	5.21	4.87	4.56	4.29
8	6.73	6.21	5.75	5.33	4.97	4.64
9	7.44	6.80	6.25	5.76	5.33	4.95
10	8.11	7.36	6.71	6.14	5.65	5.22
12	9.39	8.38	7.54	6.81	6.19	5.66
15	11.12	9.71	8.56	7.61	6.81	6.14
20	13.59	11.47	9.82	8.51	7.47	6.62

The formula used to calculate the cumulative discount factors, **Pa**, for annually recurring costs is:

$$Pa = x(1 - x^n)/(1 - x) \quad \text{where} \quad x = (1 + i)/(1 + d)$$

i = inflation rate over and above the general rate (**i** is assumed to be zero)

d = discount rate

n = number of years for which the payment is made

APPENDIX F

Non-SI and SI units useful for lighting calculations

Quantity	Unit	Standard symbol	Dimension	Conversion factor to SI unit
LIGHT				
ILLUMINANCE				
SI unit = lm /m2	lux	lx	lm / m^2	1.00
or lux (lx)	metre-candle	-	lm / m^2	1.00
	phot	-	lm / cm^2	10 000.00
	foot-candle	-	lm / ft^2	10.76
LUMINANCE				
SI unit = cd / m2	candela / metre^2	cd / m^{2}	cd / m^2	1.00
	nit (MKS system)	-	cd / m^2	1.00
	stilb (CGS system)	sb	cd / cm^2	10 000.00
	-	-	$cd / inch^2$	1550.00
	-	-	cd / ft^2	10.76
	apostilb*	asb	cd / m^2	0.318
	lambert*	-	cd / cm^2	3184
	foot-lambert*	ft-L	cd / ft^2	3.426
LUMINOUS FLUX				
SI unit = lumen	lumen	lm	$lx \times m^2$	1
ENERGY / HEAT				
SI unit = Joule	Joule	J	-	1
	kiloJoule	kJ	-	1000
	megaJoule	MJ	-	1 000 000
	calorie	cal	cal	4.18
	kilocalorie	kcal	kcal	4180
	Watthour	Wh	$W \times h$	3600
	kiloWatthour	kWh	$kW \times h$	3 600 000
	British thermal unit	BTU	-	1055.06
POWER				
SI unit = Watt	Watt	W	J / s	1
	kiloWatt	kW	kJ / s	1000
	megaWatt	MW	MJ / s	1 000 000
	kilocalorie / hour	kcal / h	-	1.16
	horsepower	hp	-	745.7

* These imperial units are for luminance of an ideal diffuser (e.g. The luminance of an ideal diffuser reflecting or emitting one lumen per square foot is the foot-lambert [ft-L]).

Quantity	Unit	Standard symbol	Dimension	Conversion factor to SI unit
LENGTH				
SI unit = metre	metre	m	-	1
	millimetre	mm	-	0.001
	micrometre	µm	-	1E - 06
	inch	in	-	0.0254
	foot	ft	-	0.035
	yard	yd	-	0.9144
CAPACITY				
SI unit = m³	cubic metre	m³	m x m x m	1
	litre	l	-	0.001
	Imperial gallon	gal (imp)	-	0.004546
	US gallon	gal (US)	-	0.003785
TEMPERATURE				
SI unit = °Kelvin	degrees Kelvin	°K	-	1
	degrees Centigrade	°C	-	*
	degrees Fahrenheit	°F	-	* *

* Temperature conversion: Centigrade / Kelvin
°K = °C + 273 or
°C = °K - 273
Example: 25°C = 273 + 25 = 298°K

* * Temperature conversion : Fahrenheit / Centigrade
°C = (°F - 32) / 1.8 or
°F = 1.8 x °C + 32
Example: 212°F = (212 - 32) / 1.8 = 100°C

www.ingramcontent.com/pod-product-compliance
Ingram Content Group UK Ltd.
Pitfield, Milton Keynes, MK11 3LW, UK
UKHW051808130526
5758IPUK00006B/101